VISIONARY
PLANT
CONSCIOUSNESS

VISIONARY PLANT CONSCIOUSNESS

The Shamanic Teachings of the Plant World

EDITED BY J. P. HARPIGNIES

Park Street Press
Rochester, Vermont

Park Street Press
One Park Street
Rochester, Vermont 05767
www.ParkStreetPress.com

Park Street Press is a division of Inner Traditions International

Library of Congress Cataloging-in-Publication Data
Visionary plant consciousness : the Shamanic teachings of the plant world / edited by J. P. Harpignies.
   p. cm.
   Summary: "23 leading experts reveal the ways that psychoactive plants allow nature's 'voice' to speak to humans and what this communication means for our future"—Provided by publisher.
   ISBN-13: 978-1-59477-147-7
   ISBN-10: 1-59477-147-2
   1. Psychotropic plants. 2. Medicinal plants. 3. Shamanism. 4. Environmental degradation—Religious aspects. I. Harpignies, J. P.
   QK99.A1V57 2007
   581.6—dc22
                                                              2006034889

Printed and bound in Canada by Transcontinental Printing

10 9 8 7 6 5 4 3 2 1

Text design and layout by Jon Desautels
This book was typeset in Sabon with Agenda as the display typeface

To send correspondence to the editor of this book send a first class letter to the editor c/o Inner Traditions • Bear & Company, One Park Street, Rochester, VT 05767, and we will forward the communication.

Contents

Part Two

PSYCHEDELICS, SCIENCE, AND
WAYS OF KNOWING

Part Three

SACRED PLANTS AND HUMAN CULTURES

Part Four

BRAZIL'S MODERN, ENTHEOGEN-BASED RELIGIONS

Foreword

Humans have long considered psychoactive plants as teachers that enhance healing, thinking, and perception. But many contemporary societies outlaw these plants.

True, working with psychoactive plants is tricky. They have a shadow side, they take time to get to know, and approaching them casually is inappropriate. But many indigenous people claim nature speaks to humans through these plants; and entertaining such a dialogue, which is the job of shamans, helps keep the human community healthy. In contrast, most contemporary North Americans and Europeans fear using shamanic plants. This is due to history, and to lack of know-how.

This book contains the spoken words of people who seek to inform Westerners about these powerful plants and about the shamans who know how to use them. Their words are clear and simple, and often based on personal experience. The result is an accessible book packed with knowledge. Ethan Nadelmann sums up its main goal: to learn to live with the world's wide range of plants so that they cause the least possible harm and bring the greatest possible benefit.

I invite readers to enjoy this book as an intellectual adventure.

JEREMY NARBY, PH.D.

Jeremy Narby presently works for the Swiss-based non-profit Nouvelle Planète to assist indigenous groups to map and defend their ancestral lands. (A more extensive biography of Jeremy Narby appears on page 206.)

Preface

The material in this book originated at the Bioneers Conferences that were held between 1990 and 2004. Bioneers is an annual gathering of environmental and social visionaries who are working *with* nature *to heal* nature. The work of these biological pioneers is focused on both practical and visionary solutions for restoring the Earth and its peoples, in recognition of the symbiotic interdependence of the web of life.

You might wonder why a forum focused on environmental restoration and social justice would address the gnarly topic of hallucinogenic plants. The answers are woven brightly throughout these pages.

At this critical moment in time, our species is straddling an unprecedented historical discontinuity in our relationship with the natural world. We are compromising the very life-support systems of the planet on which all life depends. However, the solutions to many of our most dire challenges reside precisely in the nature we are heedlessly destroying. The Bioneers, including those represented in this book, propose that there is a profound intelligence in nature, a more-than-human "mind of nature" whose operating instructions we need to learn and mimic to successfully navigate this epic cusp in human and planetary evolution.

Many ancient indigenous traditions, as well as modern re-inventions of traditional knowledge, suggest that a special intelligence resides in certain sacred plants and organisms, including cacti and fungi. This "vegetable mind," they tell us, can open new and different ways of seeing the world that reveal the most basic of biological truths: As human beings, we are part of nature. We *are* the environment.

The voices in this book believe that these sacred plants hold crucial keys to transforming our consciousness in order to reconnect us with the natural world and our own nature, to help us find our rightful and harmonious place in the web of life. That is why the Bioneers Conference has long included this controversial topic as one important piece of the larger puzzle of environmental protection and restoration.

The illustrious contributors to this collection counsel us that shamanic plants deserve great respect. Indigenous and traditional cultures honor them as profound healing and spiritual forces that people must approach with caution, sobriety, and humility. These plants are powerful teachers that can also be dangerous or harmful under certain circumstances or for certain people. Sadly, however, due to the societal misinformation and disinformation shrouding them, there is scant scientific research to help us better understand both their gifts and their perils.

For these reasons, we hope that this remarkable assemblage of some of the most brilliant, learned, and wise explorers among us will shed real light on this tragically taboo subject. If these individuals are correct, the Earth is speaking to us through these magical sacraments, making the invisible world visible, helping us read the mind of nature. Never have we needed so acutely to grasp the story of interdependent oneness, and to act accordingly.

KENNY AUSUBEL
FOUNDER, BIONEERS
SANTA FE, NEW MEXICO
JANUARY 2006

A 1972 graduate of Columbia University, Kenny Ausubel is an award-winning author, filmmaker, and social entrepreneur specializing in health and the environment. In 1990, Ausubel founded the Bioneers Conference and its parent nonprofit organization, the Collective Heritage Institute (CHI). The Bioneers Conference brings together leading social and scientific innovators focused on practical solutions for restoring the Earth. The event bridges environmental technologies with social, spiritual, and cultural solutions. Ausubel founded Bioneers shortly after co-founding Seeds of Change, Inc., the nation's leading national organic seed company promoting "backyard biodiversity." In 1999, he wrote *When Healing Becomes a Crime: The Amazing Story of the Hoxsey Cancer Clinics and the Return of Alternative Cancer Therapies* (Inner Traditions, 2000). Ausubel has also produced three documentary films: *Los Remedios: The Healing Herbs* (1982), *Hope and a Prayer: Attitudinal Healing with Dr. Bernie Siegel* (1985), and the award-winning, feature-length *Hoxsey: How Healing Becomes a Crime* (1987).

Introduction

This book is a collection of edited talks and panel discussions on the subject of the main shamanic (i.e., vision-inducing/consciousness-altering) plants of the New World and their relationship to human culture and knowledge. As such, it features a very diverse range of some of the leading figures associated with the study and/or use of the psychoactive plants of the Americas. These figures include ethnobotanists, anthropologists, medical researchers, writers, gardeners, artists, thinkers, cultural and religious figures, et al. A few of the contributors (Dr. Andrew Weil, Wade Davis, Michael Pollan) are very well-known in the larger culture for other accomplishments; several are celebrities within certain substantive subcultures or sectors of the population (Terence McKenna, Alex Grey, Paul Stamets, Jeremy Narby, John Mohawk); some are highly respected but generally known only within their scientific fields (Dr. Charles Grob, Dennis McKenna); and others are not widely known outside their own smaller circles of cognoscenti.

The names I mentioned above are predominantly white males, because the most well-known figures in these fields, those whose names are most widely recognizable, *are* white males. Yet this collection also features some strong, highly accomplished, feminine and indigenous thinkers in its mix, something very frequently lacking in other books that have covered this terrain. Their voices are not heard enough, and what they have to say is every bit as fascinating as the talks by the more famous folks with whom they share these pages.

The essays are a mix of solo presentations and panel discussions, and some of the contributors make several appearances in the book. We very intentionally wanted to offer a mix of lectures by individuals, dialogues, "trialogues" and larger panel discussions. The ethnobotanist and artist Kathleen (Kat) Harrison is the contributor who appears most frequently, since she led and convened many of the most interesting panels we witnessed over the years and especially because her unique approach to plant wisdom brings a vitally important, deeply conscious, feminine voice to the discussion.

Another unusual aspect of this collection is that we intentionally chose to include many presentations that place a special emphasis on the relevance of plant-induced visions and shamanic teachings to humanity's larger ecological crisis. That relationship is one of the main themes of this volume.

What possible interest could the use of shamanic plants present to those with very tangible concerns about environmental degradation? Well, first, while not often discussed, it is an undeniable fact that quite a few of the most dedicated environmental activists of our era were deeply affected by their experiences with psychoactive substances. In many cases, the encounters of these young people with hallucinogens in natural settings either triggered or enhanced powerful "biophilic" feelings—spiritual bonds with the natural world that set a tone for the rest of their lives. And, of course, though no one in the mainstream likes to acknowledge it, the influence of these substances on our culture's music, literature, art, fashions, mores and even technology (the cyber revolution's origins are awash in LSD), while far from being uniformly positive or free of silly excess, was/is powerful and long-lasting. So to ignore vision-inducing substances, whatever our ultimate view of them, is to ignore and distort some of our own very recent history.

The polarization around the counterculture has, if anything, increased in the decades since the counterculture emerged. This has had the unfortunate effect of stymieing the resumption of serious scientific work on hallucinogens that began in the 1950s (with a few recent exceptions, such as Dr. Charles Grob's and Dr. Rick Strassman's work). Thus, serious discussion of these substances is currently a virtual taboo in our culture; they are predominantly perceived to be either a law enforcement problem or subject to the ridicule of late-night comedians.

This is strange in light of the fact that vision-inducing plants seem to have played significant roles in some of humanity's greatest founding civilizations. The sanctuary of Eleusis in ancient Greece was among the most important spiritual institutions of that era. Its mystery teachings—which lasted nearly a millennium and were a rite of passage for nearly every major Greek intellectual figure—included the use of a mysterious vision-inducing beverage as a central sacrament. Gordon Wasson, Albert Hoffman, and others have speculated that this beverage may have been derived from the ergot mold, from whence sprung LSD. The equally mysterious use of a substance called "soma" is mentioned in India's ancient Veda texts, which are among the planet's oldest spiritual writings.

More immediately relevant, some of the indigenous groups in the Americas, well-known for the depth of their botanical knowledge and the sophistication of their ecological awareness, have long used sacred visionary plants as central tools in understanding their environment. We, who have failed so dramatically in finding a sustainable, harmonious balance with our own environment and with other species, have a great deal to learn from the practices and worldviews of these long-lived (and now highly threatened) cultures. This doesn't mean we can or should necessarily emulate their behavior or adopt their religious beliefs, but it would certainly behoove us to forgo our long-standing ethnocentric arrogance and open our minds to their ways of seeing the world. This is another principle theme of this collection.

I want to make it crystal-clear that this text is not an exhortation to the mass consumption of hallucinogenic plants. There is no denying that these are potent substances not made for casual use by the unprepared. The widespread use of vision-inducing drugs, while nowhere near as socially harmful and costly as the use of tobacco or alcohol or hard drugs, has definitely posed real problems at times. Substances that are, for the most part benign and socially constructive when used in their original cultural contexts, can become harmful when transplanted to our modern and far more disorderly, neurotic, and surreal mental landscape.

There are no facile answers as to how to best approach these powerful tools of consciousness exploration and, indeed, many of the contributors to this volume sound a significant cautionary note about their use. These folks are serious, sober, hard-working, critically minded, thoughtful people who have rigorously investigated sacred plant use in a wide range of ways. They are devoting their lives—with great courage and passion—to exploring these fascinating but tricky realms. And while many of them do have their roots in the much-maligned counterculture that emerged in the 1960s, they are the very antithesis of the clichéd image of the tie-dyed space cadet.

We don't all need to imbibe ayahuasca or psilocybin or peyote to appreciate the sophistication of indigenous ecological wisdom, but some of the most profound teachings on our species' relationship to the web of life have emerged from shamans in those cultures for whom visionary plants are a central method of knowing, of "reading the mind of nature."

So, the least those of us who profess to care deeply about the fate of the biosphere can do is listen attentively to those who have, at often great personal and professional risk, immersed themselves in attempts to understand what these cultures and these plants might teach us about how to awaken our own civilization from its current ecocidal trajectory.

The first part of the book, "The Western Mind's Encounter with Indigenous Worldviews and Visionary Plants," addresses the radical differences between ancient, shamanic, animistic worldviews and Western ideologies, and includes the views of some brilliant modern thinkers who have wrestled with the implications of this divergence. The opening chapter is a talk by the Swiss-based anthropologist and indigenous land-rights advocate Jeremy Narby, who is perhaps the most daring legitimate contemporary explorer of some possible convergences between shamanic and scientific ways of seeing the world. In the first chapter in this first part of the book, Jeremy describes his disconcerting initial encounter with shamanic wisdom in the Peruvian Amazon and his subsequent efforts to integrate his experiences.

In chapter 2, "Shamans Through Time: Tricksters, Healers, Voodoo Priests, and Anthropologists," Jeremy was joined by the late renowned Iroquois journalist, activist, intellectual, and author John Mohawk—and one of the "wild men" of modern anthropology, the irrepressible Francis Huxley (Aldous Huxley's nephew). As a jumping off point, they use Jeremy and Francis' seminal book, *Shamans Through Time, 500 Years on the Path to Knowledge,* an eye-opening compendium of accounts written over a five-hundred-year period, accounts that mark changing Western attitudes toward indigenous spirituality. These three raconteurs regale the audience with startling anecdotes and take their discussion into highly unexpected domains as they take us on a wonderful trip that includes difficult questions, highly entertaining anecdotes, and flashes of deep insight.

Chapter 3, "Culture, Anthropology, and Sacred Plants," introduces the astonishingly eloquent and deeply erudite Wade Davis, one of the planet's most passionate defenders of nomadic peoples and cultural and biological diversity, and one of its best storytellers as well. In this important discussion, Wade emphasizes the crucial importance of cultures as carriers of unique ways of interpreting the world and looks at how plant use varies among these cultures.

Part 2, "Psychedelics, Science, and Ways of Knowing," opens with a 1993 talk by one of the most extraordinary personalities ever associated with visionary plants, Terence McKenna. Perhaps the most gifted orator of our era, he was a fascinatingly paradoxical figure: an absolutely charming but somewhat misanthropic mystic, a blindingly erudite genius who never achieved mainstream recognition, and a down-to-earth guy who advanced a range of astonishing, prophetic scenarios. He brought verve and excitement to this field of study and, since his tragic death in 2000, things have never been quite the same. Quite simply put, there will never be another like him.

Here he delivers what is, in part, an uncharacteristic talk, in that it deals at length with our planetary environmental impasse. However, in its final passages, Terence offers, in his inimitable voice, what may be the signature theme of this whole book: visionary plants offer those who are properly attuned a visceral way to communicate with the intelligence of the natural world, and there has never been a time when hearing what that intelligence is trying to communicate has been more crucial.

In the next chapter Terence joins his brother Dennis—one of the nation's leading ethnopharmacologists—and Wade Davis, for an unforgettable round-table. This is a remarkable panel on several counts. First, these are three of the most brilliant thinkers ever to seriously study (albeit in very different ways) sacred plant use. Second, Terence and Dennis hadn't seen Wade since they had run into him in the Amazon during a wild expedition twelve years earlier, so this marked a fascinating reunion. Finally, they presented very different views on visionary plant use. The contrast between Terence's gnostic and mystical tendencies and his suspicion of scientific reductionism, Dennis's highly interesting defense of science, and Wade's insistence on the primacy of each culture's unique genius, made this debate incredibly stimulating and, within this field, a truly historic event.

Chapter 6, "The World Spirit Awaits Its Portrait: True Tales from One of the Planet's Great Visionary Artists," is a fitting final chapter for this section of the book. In it, one of the world's most beloved visionary and sacred artists, Alex Grey, who has been deeply courageous in acknowledging his debt to psychedelics of all stripes in helping him access deeper realities, takes us on a journey through a whole period of his career. This journey is replete with fascinating stories of his performance antics, his evolving quest for universal spiritual truths, and the emergence of his most famous body of work, the extraordinary *Sacred*

Mirror series, one of the most important artistic achievements of our generation. Alex offers a living example of the successful integration of visionary information into an exalted human enterprise in our own culture.

The third part of the book, "Sacred Plants and Human Cultures" is the longest section and really enters into the heart of the relationship between humans and visionary plants. Chapter 7 opens with a great discussion between Wade Davis and Michael Pollan—the investigative journalist, professor, and best-selling author (*The Botany of Desire,* etc.). The contrast between Michael's love of gardens and Wade's attraction to nomadic cultures is only one striking feature of a profoundly interesting dialogue about our species' relationship to plants.

In chapter 8, "Women, Plants, and Culture," Kathleen Harrison describes the special relationship that has developed between women and plants throughout most of the human journey and among most cultures. She does so in her uniquely captivating way and with a startling depth of insight and a gentle yet fierce feminine sensitivity.

The panel that discusses the topic of chapter 9, "Visionary Plants Across Cultures," must be one of the most diverse panels ever assembled to discuss shamanic plant use. This 1992 gathering brought together the world-famous pioneer of integrative medicine, Dr. Andrew Weil, who also has a long-term interest in sacred plant use and who studied botany at Harvard under the legendary Richard Evans Schultes; Dr. Charles Grob, one of the few medical researchers now legally pursuing studies on hallucinogens and a pioneer in that field; Marcellus Bear Heart Williams, a medicine man and representative of the peyote-using Native American Church; Dr. Edison Saraiva and Florencio Siquera de Carvalho, two of the first representatives to visit the United States from the Brazilian União do Vegetal Church (which uses ayahuasca as a sacrament); and Kathleen Harrison and Dennis McKenna. This chapter also includes an update on the research on psychedelics carried out by Dr. Charles Grob, Dennis McKenna and others from 1992 up until the present time.

In chapter 10, Kathleen Harrison and Paul Stamets—the ground-breaking mycologist, author, and multi-faceted researcher of and thinker about the fungal realm—engage in a very important exploration of the risks and rewards of using visionary plants to explore consciousness within the context of our culture, and tell some great stories as well. Paul is a renowned mycologist, but that moniker doesn't do justice to his creative brilliance or to the extraordinary scope of his work. One is forced to seek to coin new terms to attempt to describe him: myco-technologist, rhyzomic entrepreneur, mycelial mystic, apostle of the "fungal Internet" . . . His recent work on the bioremediative potential of certain mushrooms species to clean up lethal toxins in the environment and on natural fungal pesticides is mind-bendingly promising.

In chapter 11, "Plant Spirit," Kat and Paul join a discrete but deeply admired "plant-woman" with a legendary "green thumb"—Jane Straight—and the inspired, mysterious poet and altered-state researcher, Dale Pendell, author of some of the most astonishingly massive, deeply researched, impossible-to-describe books on plant lore. The resulting fascinating conversation took truly unexpected directions.

In chapter 12, the heroic Mohawk midwife and environmental health researcher Katsi Cook radically departs from the tone of most of this book as she immerses us in an indigenous worldview of people's relationships to plants and the natural world. In addition to her elo-quence and the beauty of some of the stories she tells, this may be among the most important pieces in this collection in that it forces us to viscerally comprehend how radically different a native worldview is from our own, and how even the most open-minded of us can be shock-ingly unaware and insensitive.

Katsi's talk is also fascinating in its description of pan-tribal cultural influences among contemporary indigenous peoples in many domains, but also as it relates to the modern use of peyote in a sacred context.

The two chapters in part 4 focus on the relatively recent emergence in Brazil of new, syncretic religions that use *ayahusca* (also referred to as *hoasca* and *vegetal*) as a sacrament. (*Ayahuasca* is a hallucinogenic plant concoction, which has been used for centuries in South America for healing and divination.) In chapter 13, the anthropologist and scholar Luis Eduardo Luna, a pioneer in helping bring ayahuasca-inspired visionary art to the larger world and who has done the most thorough work on the topic of Brazil's hoasca-based religions, explains these emerging churches and traces their evolution.

In chapter 14, "The Extraordinary Case of the United States Versus the União do Vegetal Church," Jeffrey Bronfman, a leader of a church that is an American branch of the União do Vegetal (UDV), describes the incredible legal odyssey he and his cohorts were plunged into as a result of aggressive U.S. government repression of their church. This remarkable case, which in February of 2006 resulted in an historic, unanimous U.S. Supreme Court ruling in favor of the UDV, addresses fundamental questions of religious liberty and has enormous implications for the future.

Finally, we have the Epilogue, "The Madness of the War on Drugs— A Tragically Flawed Policy's Ecological and Social Harms." The reality that it reflects is an unfortunate one, and yet sadly no discussion of psychoactive plant use is complete without acknowledging the broader socio-political and cultural context. That context is a climate of ignorance surrounding visionary plant use and an intolerance toward the fundamental human impulse to explore altered states of awareness.

Offering a systematic, devastatingly thorough indictment of the insanely destructive "war on drugs," it features Michael Stewartt and Ethan Nadelmann. Michael is an internationally recognized leader in environmental restoration who previously founded the unique and highly influential "eco air force," the airborne environmental education organization Lighthawk. Ethan is one of the world's leading scholars on drug policy and the Executive Director of the Drug Policy Alliance, the nation's leading drug policy reform organization. While Ethan has relentlessly studied the facts and figures relating to the massive social harms resulting from our drug policies and lobbied incessantly across the country, Michael has witnessed firsthand the environmental degradation that the war on drugs has caused across the continent from above.

Together they paint a picture of one of the most tragically misguided policies in human history.

J. P. HARPIGNIES

J. P. Harpignies is an associate producer of the renowned annual Bioneers Conference (www.bioneers.org) and a co-founder of the Eco-Metropolis conference (www.ecometropolis.org) in New York City. A former program director at the New York Open Center, he is the author of two books: *Political Ecosystems* and *Double Helix Hubris,* as well as the associate editor of *Ecological Medicine* and *Nature's Operating Instructions,* the first two books in the Bioneers' book series. J. P. has also been an instructor of Taijiquan in Brooklyn for twenty-two years.

PART ONE

The Western Mind's Encounter with Indigenous Worldviews and Visionary Plants

1

A Young Anthropologist Finds Far More than He Bargained for in the Peruvian Amazon

Jeremy Narby, Ph.D.

I strolled into the Peruvian Amazon eighteen years ago, a budding anthropologist fresh from the suburbs and the library. I had no previous experience with the tropical forest or its indigenous inhabitants. Back in those days development experts said that to develop the Amazon you had to confiscate the forest from its indigenous inhabitants and cut it down to exploit its resources. They said Indians didn't know how to use their resources rationally and that confiscating their territories was economically justified. I wanted to study how the Ashaninka people living in the center of the Peruvian Amazon used their forest to demonstrate that they used it rationally and therefore deserved the right to own their lands. The point was to contradict the international development banks and to try to bring about a change of policy.

The Ashaninka people I lived with took me under their wing and showed me what they knew about the forest. The Peruvian Amazon is the epicenter of world biodiversity, with more species of mammals, reptiles, birds, trees, amphibians, ants, butterflies, and so on, than any other place of comparable size in the world. It tops the charts of biodiversity. When you walk into this forest, it's a blur of species, yet the Ashaninka people who accompanied me into the forest had names for almost every plant and ascribed uses to about half of them. They had plants that they used as food, building materials, medicines, cosmetics, and dyes. They knew plants that accelerated the healing of wounds, cured diarrhea, or healed chronic backache. Each time an occasion arose I tried these plant remedies on myself, only to find that they worked, so I began asking my indigenous consultants how they knew what they knew about plants.

Their answer was enigmatic: they said knowledge about plants comes from plants themselves. They said their shamans take *ayahuasca,* which is a blend of hallucinogenic plants, or eat tobacco concentrate and speak, in their visions, with the essences which are common to all life forms and which are sources of information. They said nature is intelligent and speaks to people in visions and dreams. Well, as a young Marxist intellectual, I didn't take seriously what these Indians were telling me. It couldn't be true, because to accept that there's verifiable information in hallucinations is the definition of psychosis. It was an epistemological impossibility, and besides, it contradicted the main argument underlying my research, which was to demonstrate that these people used their resources rationally.

Nevertheless, after four months in this village, I found myself in a neighboring village one night drinking manioc beer with some men and once again asking them about the origins of their plant knowledge. One man said: "Brother Jeremy, if you want to know the answer to your question, you have to drink ayahuasca, and if you like, I can show you sometime." Several weeks later I found myself on the platform of a quiet house with this *ayahuasquero,* and he administered the bitter brew and started singing songs in the dark, delicate loops of slightly dissonant melodies. And I eventually found myself surrounded by enormous fluorescent serpents that were fifteen yards long and one yard high and hair-raisingly real. They started communicating to me in a kind of thought language, telling me things about myself that were painfully true, including that I was just a tiny human being who didn't understand very much.

Suddenly I could see that my ordinary materialistic perspective on reality had serious limits, starting with its presupposition that what my eyes were showing me didn't exist. My worldview began to collapse in front of me, and then I had to vomit. The word in Ashaninka for ayahuasca basically translates as "snake vomit." I stood up and stepped over these snakes and started vomiting colors, and I was able to see perfectly in the dark. Then I flew out of my body, miles above the planet, until the shaman shifted his song and I landed back in my body. I saw hundreds of thousands of flashing images, and one that stood out in particular: the veins of a human hand flashing back and forth with the veins of a green leaf, telling me it was all the same stuff.

The next day I tried to come to terms with this experience. On one level it confirmed what my Ashaninka friend had said: this was the television of the forest; you see images and learn. I had learned that I was puny and somehow part of nature. I went down to the river to freshen up and took a green leaf and held it up to the sky and compared it to the skin of my hand and saw that it was simply true. I felt like a mobile plant, reconciled with nature, just a human being perhaps, but proud to be a human being.

The overall experience was an antidote to the anthropocentric gaze of anthropology, but it was also too much. How could I begin talking to my colleagues about this and hope to have them take me seriously? So I chickened out. I turned my back on this mystery, and continued another twelve months of field work, studying the rational uses the Ashaninka made of their resources. I then returned to university, wrote a dissertation, and became a "doctor in anthropology," sort of like getting a driver's license.

Subsequently, I got a job working for an NGO promoting indigenous land-titling in the Amazon. This work has allowed the demarcation of about nine million acres in the name of indigenous communities, about 1 percent of the total Amazon rain forest. Over the years, my job has enabled me to meet people from dozens of indigenous societies. This has been a great privilege. During these trips I would always ask them how they knew what they knew about plants, and they all gave roughly the same answer. They all said knowledge about plants comes from *ayahuasqueros* and *tabaqueros,* who take their plant mixtures and speak in their visions with the essences which are common to all life forms. Okay, but what did that mean, really?

This was a mystery. Here are people living in the most biologically diverse place on Earth, their extraordinary knowledge about plants is now widely recognized by science and industry, and yet they say that a good part of this knowledge comes from the hallucinations of their shamans. What could this mean? I went to Rio in 1992 to the Earth Summit and found that everybody was talking about the ecological knowledge of the indigenous people and in particular the Amazonian people. And yet nobody was talking about the hallucinatory origin of part of this knowledge as the indigenous people themselves discuss it, so I decided to look into this question.

After months of thinking and reading, I began to see correspondences between what indigenous shamans say about the essences common to all life forms and DNA, the informational molecule at the heart of nearly every cell of every living being. I found that one could co-map the descriptions offered by shamans and by molecular biologists and gain a deeper understanding of both worldviews. By combining the perspectives of science and shamanism, I came to see an intelligence in nature. This is a concept that shamans have long suggested and that scientists now confirm in their own way in their recent studies of even the simplest of organisms.

Take the slime mold, *Physarum polycephalum.* This brainless, bloblike single-celled organism usually behaves like a glistening mass of mucus, which swarms over and engulfs its food. Not exactly a prime candidate for intellectual achievement, this slime is nonetheless capable of consistently solving mazes. It's a particular organism in that it's a single cell but it can grow as big as a human hand, and it can also rejoin itself if cut up. When pieces of slime are placed in a maze, they initially spread out and form a single organism that occupies all the corridors of the maze, but when food is placed at the start and end points of the maze, the slime reduces its body down to a tube which spans the shortest distance between food sources, and it solves the maze in this way, each time it is tested. The common view is that intelligence requires a brain and brains are made of cells, but in this case, a single cell behaves as if it had a brain.

The word "intelligence" comes from the Latin verb for "to choose between," and implies the capacity to make a decision. The cells in our bodies are constantly making decisions and responding to a wide variety of electrical, chemical, tactile factors so as to grow and differentiate in a coordinated way. Cells communicate with one another along signaling pathways which include domino-like cascades of proteins and a wide variety of signals for things such as: "Stay alive, kill yourself, release this molecule you've been storing, divide, don't divide, etc." Any given cell receives hundreds of such signals at any one time and has to integrate them and decide what to do.

Scientists have found recently that ants can cultivate mushroom gardens with antibiotics, bees can handle abstract concepts with brains the size of a grain of sugar, crows can build standardized hook-like tools, dolphins can recognize themselves in mirrors, and parrots can say what they mean.* The old dogma that required scientists to look at natural beings as if they were objects devoid of intention, which Jacques Monod called the cornerstone of the scientific method, no longer fits the data. Rather, there are clear signs of intelligence at all levels of nature and the concepts of indigenous shamans cast light on this.†

---

*For more information on this, see: "A Farming Ant and Its Fungus Are Ancient Cohabitants" by Natalie Angier, in the Science Times, *New York Times,* Tuesday, December 13, 1994.

†Jacques Monod was a French Nobel Prize–winning (1965) biochemist who argued in his famous book, *Chance and Necessity,* that life and evolution were the result of chance.

Even vegetables aren't stupid. Amazonian shamans have long said that certain plants could be teachers, and now even scientists are starting to recognize that plants move and react to the world with intelligence. For example, the parasitic plant called dodder moves around by wrapping itself around other plants and estimating their nutritional quality. Within an hour after initial contact, dodder decides whether to stay or whether to move on. If it stays, it takes several days before it begins to benefit from its host's resources, but dodder anticipates how fruitful its host will be by the number of coils it grows. Growing more coils allows greater exploitation, but if its host is poor in nutrients, this wastes precious resources because dodder does not have leaves and relies almost exclusively on its hosts for food and water, so it has to make right decisions or face the possibility of death. Botanists have found that dodder correctly assesses when to stay and when to move and that its foraging strategies have the same mathematical correctness as those of animal foragers. But dodder somehow computes the right decision between close alternatives without the benefit of "a brain."

Shamans use their brains to understand the world, and they have different techniques to modify their consciousness. They say this allows them to communicate with other species. They do this through a densely metaphoric language, a "twisting language," because they say the spirits of nature are fundamentally ambiguous, so frontal words crash into them. One has to spiral around them with indirect language to see them clearly. In this case metaphor is the only correct way of naming things. Shamans also say that communication with these ambiguous entities allows knowledge and power, which is itself ambiguous, double-edged and with a dark side.

And it should be said that it does not suffice to drink ayahuasca to understand the world. It requires training and preparation. It is a powerful hallucinogen, and its consumption is not without risks for casual users. For example, it can cause you to change your worldview without having bargained for it. Ayahuasca is deep water.

Indigenous people have long talked about human kinship with

nature. We now know that mushrooms, worms, giraffes, and humans have overlapping DNA sequences. Molecular biology as a whole is a demonstration of the deep kinship we have with other species. Biology waves the double helix as its flag, the symbol of the new healers, but this motif is the oldest symbol of life and healing in the world.

Shamans around the planet associate the essences or spirits with the form that historians of religion call the *axis mundi,* the axis of the world, shaped like a twisted ladder, or two vines wrapped around each other or a spiral staircase. Scientists use these exact words to describe the shape of DNA, and this shape explains its function. It's because the DNA molecule is shaped like a twisted ladder that it can be unwound; it's like two strands of a complementary text wrapped around each other that can be unwound and then copied. This shape allows DNA to be an information storage and duplication device.

Shamans say that the *axis mundi* is very long, so long that it connects the Earth and heaven. Well, if you take the DNA molecules in a human cell, you will find that they are ten atoms wide and line up to be two yards long—that's a billion times longer than its own width. It's like a little finger that stretches from Los Angeles to London. If you could take every single DNA thread out of a human body and line them all up, they would stretch 120 billion miles. How's that for an *axis mundi?*

DNA is not just an assemblage of atoms, not just deoxyribonucleic acid. It's also a kind of text. DNA communicates its information to the rest of the cell through a coding system that is strikingly similar to human codes in that the individual "letters" have no meaning. The four chemical molecules which are the rungs of the DNA ladder, and to which scientists have assigned the letters A, G, C, and T, carry no meaning individually; they have to be combined in threes for meaning to emerge. The genetic code contains 64 three-letter words, all of which have meaning, including two punctuation marks, "start" and "stop."

This kind of coding system was considered to be the proof of intelligence up until the discovery of the genetic code in the 1960s; before then, it was thought that only humans used codes in which the individual signs were meaningless. But it turns out that every cell in the world uses such a code. There's a symbolic unity underlying all of nature, and this unity isn't just limited to the genetic code; it touches every single one of our known physical chemical aspects.

I think this shows that, after five hundred years of genocide, violence, and misunderstanding, there are several ways of knowing on this planet, and that, while these ways of knowing may differ radically in their methodology, they seem to discover the same underlying truths. But despite these convergences, science and indigenous knowledge remain far apart. Indigenous people around the world find themselves in precarious circumstances. In the Amazon, they've gained land titles to extensive territories, but they're still threatened by road building, colonization, logging, petroleum extraction, and the lure of the market. Increasingly, young indigenous people in the Amazon consider nature with a market mentality, breaking with a spiritual understanding of plants and animals.

Not so long ago shamans in many indigenous societies in the Amazon used to negotiate for the release of game with the "owner of animals," an entity said to disapprove of over-hunting. But now young indigenous hunters in some parts of the Amazon have driven large mammals to extinction in response to the demand for game meat in local towns. Their market aspirations are as legitimate as anybody else's, but it will be necessary to find alternatives if nature is to be preserved in its diversity, and if the diversity of humanity is to be preserved.

So all of nature seems intelligent, and right now we certainly need all the intelligence we can get, because when humans mess with ecosystems genetic information and biodiversity go up in smoke. We don't fully understand how organisms evolved, even less about how ecosystems did. We need to understand how nature works so as to be a less harmful part of it. Indigenous people have lived in complex ecosystems for a very long time, and they say that they communicate with nature's intelligences. Conservationists might gain from working with them, and not just on "Western" terms. In biodiversity "hot-spots" such as the Amazon, the conservation of nature will require a blending of indigenous knowledge and science, but bringing these two forms of knowing together is not going to be easy.

There are methodological, conceptual, philosophical, technological, and financial differences between the two camps. For scientists and indigenous people to talk to one another they will need to develop common conceptual ground. A common ground for human knowledge would allow several ways of knowing to be compared and used together. Science has to be made more accessible in local communities so that their inhabitants can use and benefit from it, and indigenous systems of knowledge have to be made graspable and legitimate for scientists so that knowledge and wisdom may be shared.

There are promising initiatives in this direction. The people of the Peruvian Amazon have started a school of bilingual intercultural education for young people from fifteen different indigenous societies. In this school they learn the finer points of their respective mother tongue and particular culture, as well as learning Spanish and science. This program, along with others like it, needs support. A common ground for human knowledge would reconcile control and respect, microscopes and modified consciousness, scientific texts and oral expertise, detachment and emotion, matter and spirit. I found that learning to work with indigenous systems of knowledge is like learning another language. Bicognitivism, like bilingualism, is hard, but it gives you another way of looking at the world.

And there has, fortunately, been an improvement in scholars and anthropologists' willingness to begin to consider indigenous worldviews with far more seriousness and respect, because, while surprisingly little appears to have changed in shamanic practices in the last five hundred years, scholarly treatments of shamanism have changed dramatically. The biggest shift in the observers' gaze came in the 1950s when anthropologists started to apply "participant observation" to shamanism.

Participant observation consists of participating with people in their activities while observing them from a distance. It is a somewhat schizophrenic methodology, but after a while you can get good at it. When anthropologists started taking part in shamanic rituals and swallowing these bitter and hallucinogenic potions or mushrooms, they started to see that they themselves could see things in a similar way to shamans. They began to understand that shamanic phenomena pertain to the faculties of the human mind—and not just to mythology and superstition.

As a result, there have been all kinds of serious and detailed observations of shamans over the last fifty years, and in the last thirty years more texts have been written about shamans than in all previous recorded history; a multifaceted phenomenon that covers five continents was documented. Shamans emerged as people who are now respected by psychologists, epistemologists, and ethnobotanists, among others. By paying attention to what shamans say and do, scientific observers have found that they could actually learn things. I was lucky enough to be able to ride this wave right into the third millennium, but I wanted to go a little further in my budding efforts at bicognitivism, by trying to test whether two languages, those of shamanism and of molecular biology, could be brought together very tangibly.

So I'll end by describing the unusual experiment along these lines that I undertook in September 1999. I accompanied three molecular biologists to the Peruvian Amazon to work with an indigenous *ayahuasquero* to see if they could find bio-molecular information in their visions. The idea was to test the hypothesis that ayahuasca visions could reveal genuine, verifiable scientific information about DNA to trained scientists.

One of these biologists runs a genome-sequencing lab for a California genomics company. Another is a professor of molecular biology at the Centre National de Recherche Scientifique, a big French think tank. And the third is a professor of molecular biology at a Swiss University, who runs a lab where she and her colleagues modify the genes of potato and tobacco plants. Aged between thirty-nine and sixty-four, these senior scientists are serious people. They courageously availed themselves to this experience, including rigorous dieting and preparation.

They participated in three ayahuasca sessions, and they *did* see DNA molecules and chromosomes, and they *did* gain information about the research that they were conducting. The person studying the human genome works as a gene detective. As the endless sequences of the AGCT letter information come in, her job is to look for the needle in the haystack, to find the genes. (As mentioned on page 18, the letters of A, G, C, and T represent the four chemical molecules that make up DNA.) One of the techniques researchers use is to identify what they call "CpG islands": DNA sequences rich in C and G. These structures are located just upstream from about 60 percent of all human genes. If there is a CpG island, a gene is probably not far away.*

The Swiss biology professor wanted to know if all CpG islands have the same structure; she also wanted to know what their function is. In one of her ayahuasca sessions she visualized herself as a transcription protein flying above a DNA molecule, and she saw that the CpG islands all had the same fundamental structure and their function was to serve as landing pads for transcription proteins. She says that she had never thought of this before nor had she heard anybody else suggest it, and she told me last week that this remains a testable hypothesis that she is working on in her lab.

The French professor's response to the experiment was that ayahuasca is not a shortcut to the Nobel Prize. He also said he thought shamanism was a harder path to knowledge than science, at least for him, because it involves a subjective and emotional experience that is very strong. If you are a professor of molecular biology and have spent your life running labs trying to come up with reproducible experiments, it is difficult to consider learning through an experience that is subjective and non-reproducible. You can never have the same ayahuasca experience twice nor can someone else have exactly the same experience as you, so there are fundamental methodological differences between these two approaches.

---

*DNA is a "chain," i.e., a "stream" composed of four twisting molecules. Researchers look for clusters of consistent, repeating patterns (genes) in the "flow" of what seems random ("junk") DNA. In this case, 60 percent of the time that a human gene is found in the DNA flow, one of these CpG "islands" is nearby.

Nevertheless, what he had been studying for a decade with his colleagues was "How do sperm cells become fertile?" When sperm cells come out of the testes, they aren't capable of fertilizing an ovum. They have to travel through the sperm duct, or epididymis. The sperm duct secretes fifty different proteins. They work on the sperm like workers in an automobile chain, and by the time the sperm gets to the end of the duct, it can fertilize an ovum. They were doing research on this because they were looking for a male contraceptive. One of the things he wanted to know was which protein was the key to this process. He asked about this in an ayahuasca session, and a voice gave him an answer: "It's not one protein, it's all fifty and how they work together."

The third molecular biologist has been modifying the genes of potato and tobacco plants in Switzerland, where there has been a lot of opposition to genetic engineering. People like her have had their labs run over and plants torn out and she's taken these criticisms to heart. She wanted to know about the ethics of genetic engineering. Because she'd been working with tobacco, trying to make it resistant to a virus for ten years, she'd heard that shamans speak with an entity they call "the mother of tobacco," and she wanted a one-on-one interview.

She asked the ayahuasquero whether he could arrange this and he said, "No problem. " During a session he sang the melody corresponding to tobacco, and the genetic engineer reported that she indeed spoke with an entity that identified itself as the "Mother of Tobacco." She asked about genetic engineering and the mother of tobacco said: "Tobacco is here to serve. Tobacco is pleased to be of use to any of the beings of this planet, be it a virus or a human being. Modifying my genome is not a problem, as long as it is done with the interests of all of the above in mind."

Capiche?

This presentation took place at the Bioneers Conference in 2002.

2

Shamans Through Time: Tricksters, Healers, Voodoo Priests, and Anthropologists

Jeremy Narby, Ph.D., Francis Huxley, and John Mohawk, Ph.D.

Jeremy Narby: Shamans come from indigenous cultures, which anthropologists used to call "societies without writing." Europeans got on boats five hundred years ago and spread out into the world and started dominating people and describing them at the same time. The people who did the describing tended to be clergymen, and what they perceived—when they saw natives fasting for months and swallowing tobacco juice through funnels and wearing collars of stinging ants and claiming to speak with the spirits of nature—was "devil worship." And in the sixteenth and seventeenth centuries this was a very serious accusation because you could easily be burned at the stake back in Europe for it, but for the first two hundred years or so of their encounters with this phenomenon this is how all these Western observers described the practitioners we would now call shamans.

The European Enlightenment and an embrace of rationalism arrived

24

in the eighteenth century, and the first rationalist observers who came upon shamans no longer saw devil worshipers but rather "impostors." This was a step up for the shamans of the world. Now they were seen as just a bunch of tricksters who didn't really deserve to be burned at the stake! When we got to the nineteenth century, serious gentlemen were sitting in their clubs in London smoking their pipes, and Darwin had just demonstrated that we're linked to other species.

We now seem to have to admit that we're only evolved monkeys, but we in the Western industrial world have dominated everybody with our cannons. We're sitting on top of all this wealth, so we must be, obviously, more evolved than these savages, these "primitives" (from the Latin *primitivus*—"born first") who are still stuck in prehistory. But to understand who we are, where we came from, these "civilized" gentlemen decide we must study the *childhood* of humanity, so anthropology was born, based on the idea that other societies were inferior to ours, and that they could be studied scientifically. So, on the one hand, it's a bad start. Anthropology begins as a fundamentally racist science, but reality is a complex affair, so it's also a positive moment, because this is the first time that humanity created a discipline to study itself in all its diversity.

And fairly soon, in the miasma of a lot of bad anthropology, some shining lights began to emerge. Franz Boas, a German-American who spent a year with the Inuit in the 1880s, concluded that the Inuit were people just like us. They loved their children, just as Europeans did. Believe it or not, this was big news at the end of the nineteenth century. In 1900 Boas also said that the student of anthropology, if trying to study the mentality of another culture, must first take into consideration his or her own bias. An observer of another culture has to become aware of his or her own gaze. Boas trained a new generation of anthropologists, and they went out into the world. As a result, in the 1920s we started getting the first written reports of any detail and sensibility on shamans.

One among this new breed of anthropologists, Knud Rasmussen, spent years living with the Inuit in Greenland and recorded word by word what the shamans said. This was the first time that it became possible to learn, via text, about the lived experience of this phenomenon that we now call shamanism, in the actual words of its practitioners.

Anthropology evolved rapidly in the first decades of the twentieth century and created the method of "participant observation," but it took several decades after the invention of the method to apply it to shamanic studies. Then, when anthropologists started participating in shamanic sessions, all hell broke loose, especially when they started imbibing these strange beverages and discovered that what these natives had been saying all this time wasn't just a bunch of superstitions. One could test their statements. One could be an atheist or a materialist academic . . . and still see cosmic serpents.

Finally, by the last thirty years of the twentieth century, Western observers realized that there was something multifaceted and profound in shamanism that involved modification of human consciousness, botany, psychology, music, and performance. An explosion of texts started being written by people who had observed shamans carefully, and then shamans themselves actually started to publish books. One of the best of these is by Maria Sabina, the now-famous Mazatec healer. It's a little pearl of a book. This five-hundred-year process finally took us from the most closed and intolerant perspective to an understanding that one can gain real knowledge by dialoguing with humility and an open mind with these practitioners who lie somehow stubbornly outside the rationalist path. After five hundred years, we could finally get down to some serious work.

Francis Huxley: Anthropology is a most interesting discipline because it allows you to do just about anything with anything that has to do with human beings. After my first fieldwork in Brazil with a tribe of more or less ex-cannibals (I found cannibals very interesting for family reasons: there are many ways to eat people), I worked for a year in a mental hospital in Canada. It was one of the most appalling bits of education I've gone through. It was an old, over-crammed mental hospital, and

I learned the rudiments of the jargon with which people were labeled: schizophrenic, paranoid, hysterical, for example. I went away with a great feeling of dissatisfaction, wondering if people would be treated better in a society that did not use these labels.

I sought advice from a variety of people who had unconventional views. One was Eileen Garrett, who had been Conan Doyle's favorite medium and had set up an international parapsychology institute in New York, and she funded me to go down to Haiti to see what they did there. I learned a number of notable things during my stay. One was that the type of mental perturbation that overcomes shamans and voodooists is very similar. The insights come out in the same way. The only thing that distinguishes a possession cult from a shamanic one is that among shamans, it's the shaman who gets enlivened by the spirit, whilst in a possession cult many in the congregation go through a process of dissociation and trance. I've seen a dance floor filled with more than a dozen possessed people at the same time. But the actual central mechanism is the same in both cases.

One of the things that most impressed me there was the method of diagnosis the shamans employed. I've argued with psychiatrists in England about the stupidity of using one label to cover a whole heap of temperaments. Schizophrenia, for example, is a label covering about ten different metabolic disorders, and to use one label is therefore very misleading. It narrows one's imagination and one's capacity to heal. In Haiti they diagnose your disorder according to which *loa* (i.e., spirit/god) is either taking over what it should not or is out of place in you and trying to get back into balance. They have a very large pantheon of such *loas*. In order to heal, a voodoo priest will help guide patients into a state of possession in which they can manifest that spirit which is creating the disturbance, to show its face in public so that its issues can be resolved or it can express what it wanted to express through this person.

These pantheons come from West Africa, from Angola, from the Congo, even from Ethiopia. A diagnosis might be: "Aha, you're suffering from a misunderstanding with Sobo, that thunderous spirit who deafens you with ringing in your ears and spots before your eyes." There is a whole ritual activity by which you summon this spirit, and you sleep on the magical point of this spirit, and after a somewhat arduous initiation, it comes and takes you over and a process of healing can begin. After that you have to go and renew your contact with this spirit periodically because if you don't, you're back at square one.

So maintaining spiritual balance with all these forces is a lifetime's occupation, just as it is a lifetime's occupation to be a shaman. In many parts of the world to be labeled as a shaman fills people with horror and devastation, because it's seen as a murky, ugly, difficult profession that isolates you from the rest of your people. It's not something people tend to choose. It's rather an onerous fate thrust upon you because the spirits demand it.

I have met all sorts of unusual people I would call shamans, not only the traditional, indigenous types I encountered in Haiti and the Brazilian Amazon. For example, I met a very interesting urban shaman from Rio whose principal spirit was none other than the Emperor Nero of ancient Rome. According to this spirit, Nero was a very interesting man who employed sadism with great creativity. This spirit certainly knew about sadomasochism!

This man in Rio had a long lantern face, and when he was possessed by this Nero character, his eyebrows would go up, he would have a slight frown, his eyelids would half close and there would be a wolfish smile upon his face. The first time I saw it, I thought: "Well, there's a man who's saying: 'Come closer, or I'll bite you.'" But he had, like my friend the famous Scottish psychiatrist R. D. Laing, what in the Bible is called "the discernment of spirits."*

I became friendly with him, and for quite a time I would be attentive to everything that he was doing. He was very effective as a healer. I

---

*Charismatic Christians consider the "discernment of spirits" to be a spiritual gift that supernaturally enables a Christian believer to distinguish between holy and unholy spirits. This gift is thought to be especially necessary in instances when individuals may need to be delivered or healed of demonic possession.

felt that, compared to him, my own imagination and learning had been hindered by psychiatric ideas that are not very good for the soul. I had to relearn a whole different attitude so that I could actually look at people and see what he was looking at. I would follow him. He used to greet his audience in the form of Nero and go around looking, smiling wolfishly at everyone, shaking their hands. He was uncanny at knowing what people were up to, even when he didn't speak a word of their language, as was the case when he visited England.

He rather reminded me of the family doctor my mother had found when I was young, who one day came in from his rounds in the district for a cup of tea, opened the door, stopped, looked at my mother and said, "You've got amoebic dysentery," and so she had! This was the first time I had seen a correct diagnosis made at twenty paces, and I was very impressed. This Brazilian had a similar gift for diagnosis. There's nothing like apprenticing yourself to one of these people if you really want to understand what's happening. In fact, I can't think of any other way of doing it. You have to get their way of seeing into your own system.

This can only be done via oral and especially non-verbal transmission. Book learning is very useful, but it has a dreadful way of cutting one off from one's sense impressions, especially from the ones that one hears with. After my first fieldwork in Brazil, I came back and started leafing through my field notes, and I was horrified to discover that I had written down the meaning of what I had been told by the shamans, but not their exact phrases. I hadn't captured the lilt of their phrases, the poetics of their language.

I realized that I hadn't been actually listening to the words that were spoken, so I spent nearly eight months learning how to remember *words* rather than meanings, and this proved to be one of the most mind-clearing things I've done in my life. In fact, it's what has allowed me to continue my anthropological career, as far as I've been able. So, if any of you are interested in the shamanic vocation, I advise you to give up reading for about six months to start to listen to the words you're actually told, and then remember them by writing them down exactly. This will be very hard for you to do at first, but very well worthwhile afterward.

Then one must begin to observe what is conveyed non-verbally, by body language. That's really where one can discern the *spirit* of people. By learning to intuit information from the appearance of things, hidden realities make themselves known to you, and then you can begin to diagnose or understand.

John Mohawk: I come from a culture that doesn't have major psychotropic drugs, but our healers have very similar theories and sometimes use methods very similar to those of the shamans Jeremy and Francis have studied. Shamanism is a complex topic because you can find very similar symbols and approaches across cultures, but different cultures can also have very different beliefs about why certain people may have healing powers or how to develop them or how to use them. It's a good thing there's more tolerance about and interest in indigenous healing, but some non-indigenous people get so enthusiastic about it they don't realize the seriousness of it.

Among many peoples, including mine, it's almost a disaster to be tagged as a medicine person. It's really not something you wish for or wish for your kids. It's a fate that chooses you. You lose a lot of freedom in your life. You have to walk a very strict path, and you have a lot of burdens. People come to see you and expect help with really serious problems at all hours. And in the Indian communities I know, these types of healers and people with special gifts have to maintain certain rules. You can't ever brag about it or go put a shaman shingle outside. The people in your community won't use you anymore if you do that.

And issues of power and misuse of power can come up. Indigenous cultures that have a five- or ten-thousand-year experience in a given environment of plants and animals often lived in small, intimate bands, isolated from other people. Some people with special abilities might emerge in those situations and gain a certain power in the group. That can become a problem at times in any group, and even more so today when that same type of pattern emerges in our more unstable modern society, as we saw with Jim Jones's cult in Guyana.

One funny thing about culture is that what would seem to be strange and unusual ceremonies and practices to people outside a group seem perfectly normal to those who are a part of it. I remember when I was about nine or ten, living on the reservation, I would go to town a lot with older kids and go to the movies, which in those days cost thirty-five cents.

I remember very vividly one movie that was part of a sensationalistic series about "the strange world" of this and that. It was just a series of clips from around the world showing people in "exotic" cultures participating in "strange" rituals: in one scene people somewhere in India put pins through their cheeks; in the next a guy in Central America jumped off a very high platform with his feet tied to a rope; in another, folks in the South Pacific walked on burning coals; and so on and so forth. I was sitting there like everybody else, enjoying myself watching all the wacky people doing zany stuff, but then all of a sudden they showed a scene from my own culture, an Iroquois healing practice in which the healer takes hot coals and puts his hands in the hot coals and blows the disease away from somebody using coals and tobacco, and I thought, *Why is that in this movie? What's strange about it?* To me that was a normal thing I had grown up with.

Many years later, when I was the editor of *Akwesasne Notes* we went all over the country, representing the traditional movement at colleges and Indian communities and doing activist work. We were welcomed on many reservations and got to see a lot of ceremonies that weren't open to the public. And some things happened that I couldn't explain. One episode wasn't necessarily the most dramatic but the memory has stayed with me because it made me accept that some folks have special gifts. I wasn't really skeptical about that because my culture accepts that as a reality, but this one event drove it home.

At one point there was a very active movement in Cree country, and a group of militant Crees were occupying a camp in a national forest, and we went to visit them. And a very famous shaman up there invited us into their sweat lodge. He made a lot of mysterious things happen: visions and strange sensations and feelings. At one point, he was sitting there and he just started telling us exactly where we'd been, what had happened to us so far, and where we were going to go next. And these people didn't know we were coming, didn't have phones there, and we hadn't told anyone any of this information. Then he said: "You have a friend who is driving a blue bus. Today your friend is over the water somewhere, and the bus broke down, but he's going to keep going . . ." And he proceeded to tell us where he would go and what would happen to him, and so on.

We did know a friend with a blue bus at home, so when we got back I contacted the guy and asked if anything had happened to him recently. He said he had taken a trip to a Hopi snake dance, but when he got to the Missouri River, right in the middle of the bridge, his blue bus broke down. And sure enough, it had happened on the same day we were with the shaman, and everything else my friend told me about his trip was exactly what that shaman had said would happen. Later I met quite a few other people with those types of gifts, but that shaman up there in Cree country was really good. Over the years whenever I would ask one of these folks how to explain these things, they always told me not to try to explain it, that it was a mystery we can't understand, and beyond our capacity.

MODERATOR'S QUESTIONS AND PANELISTS' RESPONSES

Moderator: Jeremy, what do you think of the use of traditional psychoactive plants by non-indigenous people?

Jeremy Narby: That's a difficult question, but ultimately I think that grown-ups should be free to ingest the plants they want to.

Francis Huxley: Could you give me your definition of a grown-up?

Jeremy Narby: Good point. Well, I was going to add that I think that people should be as fully informed as possible, and so maybe that's part of the definition of what a grown-up is: somebody who informs herself. And there are plants, such as datura for example, that can have a very dark energy, really rough and potentially dangerous plants that only the most skillful *maestros* (masters) should ever use. There are *maestros* in the Peruvian Amazon who work with datura but its use is surrounded with an aura of darkness in most cases.

As to ayahuasca use, even if the idea that people shouldn't take ayahuasca out of its indigenous context and outside of the Amazon had merit, at this point, the cat is out of the bag, and many circles of people are using it, including the Brazilian religious groups that have popularized its use as a sacrament, such as Santo Daimé and União do Vegetal. I don't have any experience of these organized ayahuasca cults. I'm an agnostic, personally.

It's clear that ayahuasca is a powerful psychotropic, and that even though it's more or less brain-compatible because its main active ingredient seems to be the dimethyl-tryptamine (DMT) that mammal brains also seem to produce, nevertheless it's a very powerful mind-modifying substance, so I think that just drinking powerful ayahuasca by yourself is a recipe for trouble. It's not something that I'd recommend widely, and it's certainly not anything I practice. My practice is going to the Peruvian Amazon (which happens to be part of my job, so it's kind of handy) and working with *maestros* who have a good reputation and whom I respect and who have a clean heart.

And half the experience is the singing of the maestro. The experience is not just about guzzling ayahuasca: it's about whom you drink it with and what you drink it for. It's something that one does to explore very important questions. That said, I think it can be very helpful, obviously, in some cases for well-prepared individuals who want to do personal exploration, enhance their creativity and so on, or just achieve some physical healing.

Much has been written on this subject, but I think that the indigenous people of the Western Amazon should be deferred to as the ultimate authorities on ayahuasca, in the same way that if you wanted to talk about red wine with real experts, Bordeaux would be the place to go (with all due respect to Napa Valley). The ideal would be if people who are authentically interested in ayahuasca could go visit a sort of university created and run by the indigenous people of the Western Amazon. It's a serious path of knowledge. It's not really ideal for experimentation in the suburbs.

There's also a whole dark side to ayahuasca that's not discussed by most of the very idealistic non-indigenous enthusiasts of its use. Abuses of power and magical power struggles are integral parts of the shamanic world. The more a healer can actually heal and wield power, the stronger the temptation to abuse that power grows. That's one reason shamans are feared, and communities keep a wary eye on the healer to make sure that jaguar's dark side stays down. There's a whole world of paranoia, suspicion, witchcraft, and magic darts flying all over the place, that goes on in these ayahuasca-using societies. Ayahuasca is not presented among these groups as a groovy thing you might do next weekend. It's always seen as a kind of a shady practice that one approaches if one has to, but with apprehension.

But when we start seeing ayahuasca usage in the Western world, suddenly one only hears about healing and the positive side of it, so some discussion of the darker side of this phenomenon is called for. You can actually harm people if you know how to do it by administering ayahuasca to them and then messing with their minds. It's pretty clear that ayahuasca can fuel a thirst for power, and so one wants to keep an eye on these practitioners, especially when they get good, just as one needs to keep an eye on anyone in a position of power.

Moderator: Francis, I have a feeling you might have something to say about the dark side?

Francis Huxley: Yes. There are people I would not wish to take any mind-altering substance with, or if they were anywhere nearby. Early in my life I had very bad experiences. In fact, my very first trip was with a man who was so hyper-manic it was a painful experience. In the 1950s LSD was still in the hands of psychologists and psychiatrists, and they used to administer it in white coats with clipboards and pencils, watching you intently and taking notes, and they noticed that every one of their patients developed paranoid symptoms. I wonder why. That was certainly a form of black magic.

Shamanic practices are now often made to sound so benign, but it's a strange world of weird magic substances and darts, of sleight of hand and of inexplicable powers. That Brazilian I mentioned could manifest a magic substance he called "ectoplasm." One day, as he was working with a group doing healings, he took me aside and said: "When I get going, that man over there is going to faint, and then that woman, and I'll get their ectoplasm."

Sure enough, within twenty minutes the man fainted, and four minutes later the woman fainted. He passed a razor blade just over the ribcage of the woman who had been biopsied as having a cancer low in her left lung. A little blood oozed from the cut; he put a glass tumbler over it, and some cotton wool over that, and pressed it firmly down. After some minutes he relented, whipped off the cotton wool, and there under the glass was something white growing out of the cut. Giving us a moment to admire this prodigy, he removed the glass, detached this something—it looked like a miniature dog turd—and broke it in half to let us see a dense pink that was staining out from its core.

I had to believe what I saw, but couldn't believe that such a thing could be: the disjunction made me feel sick, sick enough that I soon went to the bathroom to throw cold water on my face. He was walking down the passage when I came out. "Eh, anthropologist, what do you make of that?" he asked.

"I don't know," I said.

"Neither do I," he said happily and went on his way.

I have no idea what manner of conjuring was involved in such a moment. Two theatrical magicians I was to meet later that year likewise had no idea how to stage ectoplasm growing out of a cut. But the effect on the patient was undeniable: I met her some years later. She was freed from her pain and anxieties and successful at her job even though she was smoking two packs of cigarettes a day.

As undeniable was that the two people the Brazilian had singled out as fainters before the event had fainted on cue, so that the ectoplasm he was talking about came from him stealing their awareness and using it to perform his cure. Others, I realized, did much the same to satisfy themselves only: one needs to be careful around such cannibals of ectoplasm.

John Mohawk: In my opinion the biggest difference between modern Westerners and indigenous peoples (at least the ones of the Western Hemisphere that I know), is their view of the sacred. For example, for the Maya, corn is a goddess and a symbol of the cosmos. Corn is pretty remarkable stuff. You can take this little grain and put it in the ground, and soon you've got an eight-foot-tall plant, and it has enough food on one cob to keep an adult person alive for one day. That's pretty good magic.

For a modern industrialized culture, food is just a commodity, but for an indigenous perspective all of life is sacred. Everything alive is imbued with a force, and it's not a force that you can take lightly. The Maya felt that since corn sustained them, they were, in a sense, made of the corn, which was sacred. They saw the connection between divinity, corn, the Earth and human flesh, and understood they were part of that cycle and that you had to respect the integrity of the spirit of the corn to continue to survive.

Now along come the Europeans. They get hold of corn, and for them it's just a product they can grow to make money. There's no sense of sacred connection to it, so later, when the era of industrial agriculture arrives, nothing stops them from using petrochemicals, herbicides, and the like to raise millions of acres of corn.

If you took all the corn raised in the United States this year, and you put it into one field, that one field would be the size of New York State. And most of those cornfields are in the watershed that drains into the

Mississippi River, so the run-off of all the fertilizers, pesticides, herbicides and everything else they use to grow that corn (and other crops) flows into the Mississippi and then into the Gulf of Mexico! And, guess what, there's a huge and growing "dead zone" in the Gulf of Mexico. Their lack of a sacred relationship to the corn eventually led to a destructive pattern, an imbalance in the cycle.

And this applies to relationships with psychoactive plants as well. Our relationships with all plants and all living things can have unintended consequences. Indian societies have long known about the risks of unintended consequences, and are very aware that there's always a dark side to power and that you have to treat these things very carefully and very respectfully. For indigenous people there's this sense that our relationships to the things of life have to be thought through and honored with ceremonies and sacrifices to make sure the right balance is maintained. But modern society has no limits or guidelines about right relationship to the other living things that sustain us.

Today corn provides more calories than any other crop in the United States, but most of those calories are in the form of corn syrup, through stuff like soda pop and sweeteners for pre-packaged foods created by the multinational corporations. These corn products are giving many of us diabetes and making a lot of us fat and sick. Corn can be an incredible food, but, misused, corn can be something very, very negative and dangerous.

Thinking about corn as a sacred food and thinking about corn as a way to make money are two very different ways of looking at the world. The modern way is thoroughly profane: it looks at corn as a mere object that we can manipulate at will. There's no sense that it has its own energy, its own will. We think we can control all other living things, and that's how you eventually lose control. And this holds true for psychoactive shamanic plants as well. You have to approach them with a lot of respect. European-based cultures can get very weird about plants. I think one of the strangest aspects of Western culture is its long oppression of people who use plants for healing.

This had been brewing for a few centuries before the Inquisition, but it really took off after the Inquisition. Let's really step back and think about this for a moment: a culture that makes the possession and use of certain plants illegal strikes me as truly peculiar. If an alien investigator landed and asked me what the strangest thing we did on this planet was, I might very well answer: "Well, we make plants illegal."

Throughout history, rulers have worried that mind-altering plants might undermine the power of the state, but we're living now in a bizarre society in which it's legal to own fifty-caliber machine guns but possession of a marijuana plant will get you in a lot of hot water. The Inquisition never completely ended.

_____

This discussion took place at the Bioneers Conference in 2002.

3

Culture, Anthropology, and Sacred Plants

Wade Davis, Ph.D.

I now have a position at the National Geographic Society as "Explorer in Residence." I realize that sounds like an oxymoronic sinecure, but if you want some evidence that social change does happen, it may be a noteworthy fact that someone who studied ayahuasca, shamanism, and zombies has an official position at the National Geographic Society. It's all part of an amazing transformation of the National Geographic Society where, after a hundred years, they realized that the magazine, successful as it had been, had been the tail wagging the dog, and the dog was *a* mission, but what was *the* mission?

A new CEO came in and, with great imagination, declared that the Society, having told you for the first century of its existence about the world, was now going to get together for the next century and help you save the world. So the National Geographic, amazing as it may sound, has absolutely embraced conservation in every dimension. And my particular bailiwick is the ethnosphere, which we're defining as the sum total of all thoughts, dreams, beliefs, and intuitions brought into being by the human imagination since the dawn of consciousness.

And, of course, this legacy of the planet's cultural diversity is severely imperiled, and we've seen the terrible "blowback" this can cause. The triumph of secular materialism is a conceit of modernity, and we have this kind of illusion that there's an inexorable trajectory of progress, and that even if we are sympathetic to indigenous and traditional societies, they are sort of quaint and colorful but on the margins of history as the real world moves inexorably forward. But one lesson of anthropology is that there is no trajectory of progress. When you look at the circumstances of indigenous peoples throughout the world, half of humanity, half of the legacy of our species, is disappearing before our eyes. And it's *our* doing.

These are not cultures pre-ordained by some impersonal destiny to fade away but are dynamic and vibrant cultures that have some insights we could desperately use. However, they are being driven out of existence by external forces beyond their capacity to come to terms with. We induce nomads off the land in Kenya and they find themselves living in the slums of Nairobi where the unemployment rate is 60 percent for those with a high school education. Or, in Malaysia, through egregious violations of their homeland, the Penan are forced into settlements. Lima, Peru, that had maybe four hundred thousand people in 1940 now is home to nine million people. Whether it's the refugee camps of the Afghan frontier or the barrios of Lima, all these places become breeding grounds for resentment.

And whenever people are pressured in this kind of way and their cultures destroyed, strange things happen. In anthropology we call them "millenarian cults." In the case of the Sioux we saw the Ghost Dance phenomenon. When the buffalo had been swept from the prairie and the people were reduced to living on reservations, a myth arose that somehow if you just prayed in the proper way the whites would be swept away, the buffalo would come back, and even the bullets of the cavalry wouldn't pierce your Ghost Dance shirt. The Boxer Rebellion, the cargo cults of the South Pacific after World War II, and some aspects of the Mau Mau Rebellion were similar phenomena. We're seeing it again in some parts of the world, with this fanatical nostalgic fantasy for an era of Islamic culture that never existed, with terrible consequences.

As I began pondering the fate of the ethnosphere and of imperiled cultures, I was drawn back to my academic roots. I began as an anthropologist. As an anthropologist working in the field I often found that botany was an ideal conduit to cultures, a great vehicle for breaking down the inherent barrier that existed, by definition, between the people with whom I found myself living as a guest and me.

I often turned to botany as a way to understand the peoples of the Amazon. I lived, for example, amongst the Barasana people who cognitively don't distinguish the color blue from the color green because the canopy of the heavens is, for them, equated to the canopy of the forest. The obvious conduit to many cultures was the botanical realm, and particularly, sacred plants. Of course, when I was working in the Arctic or in Tibet or Haiti or other places in the world where sacred plants are not prominent, I often had to find other means to seek the heart of a culture.

It's very important to realize that where sacred plants are found, their use is firmly rooted in culture. One of the most interesting aspects of these plants, perhaps 120 of which have been found in nature, is their distribution. Ninety percent of them are found in the Western Hemisphere, or, to a lesser extent in northeastern Siberia. The Old World is notably lacking in psychotropic plants, which is curious because, of course, the forests of equatorial West Africa or Southeast Asia are as botanically rich in pharmacologically active compounds as the Amazon. The peoples of those regions have explored their forests with the same dexterity and perspicacity as the people in the Amazon. And certainly, in equatorial West Africa, that seeking has yielded any number of interesting bioactive compounds. Indeed, in equatorial West Africa, the variety of folk poisons is extraordinary. And yet, with the exception of iboga, not that many psychotropic substances are actually used in Africa.

On the other hand, in the Amazon where there is a plethora of psychoactive sacred plants and they are used extensively, the use of poison is also found (as part of the hunting technology), especially in the famous blend that yields curare, a blend composed of ninety plant species or more. However, despite this very particular function of poison in the Amazon, Indian peoples of the Americas in general, from the Arctic homeland of the Inuit to Tierra del Fuego, did not use poisons on each other. So the uses of plants are firmly rooted in specific cultures.

When I later worked in Haiti, a voodoo priest told me: "You know, you white people go to church and speak about god. We hear that these Indians eat their magic plants and speak to god. Here, we dance in the temple and become god." Sacred plants are one avenue to satisfying a common and ubiquitous desire in the human spirit, the desire to periodically change consciousness, which is found in every culture. What that Haitian priest was telling me was that through the spirit of the dance, through the idea of transformation and metamorphosis inherent in spirit possession, people of his part of the world had a different avenue to seek a direct dialogue with the godhead.

One of the most interesting aspects of psychoactive plant use in a place like the Amazon is not simply its stunning pharmacological and socio and psycho-spiritual consequences but what it reveals about the mysterious genius of the shamans. In 1981 when I traveled with Terence and Dennis McKenna in the Amazon, they introduced me to the curious mystery of ayahuasca's origins.

In the Amazon you have a flora of some eighty-thousand species, and shamans somehow figured out a complex procedure to combine two completely morphologically distinct plants: a nondescript mildly psychotropic liana and a run-of-the-mill shrub. It turns out that the liana contains the compounds harmine and harmaline, and the shrub, *Psychotria viridis*, is full of dimethyl-tryptamine (the potent hallucinogen DMT) that is normally orally inactive because it's denatured by an enzyme found in the human gut, monoamine oxidase (MAO). It is only

potentiated if taken in conjunction with some MAO-inhibitor that dena-
tures or inhibits the MAO in the human gut.

Well, it turns out, of course, that the beta-carbolines found in the
Banisteriopsis caapi liana are exactly the kind of MAO-inhibitors necessary to potentiate the tryptamines in the *psychotria* shrub. How in a
flora of eighty thousand species did these shamans figure out to combine two morphologically distinct entities in different families of plants
and bearing no phylogenetic relationships to each other? How did they
know that when these two entities were combined in a quite sophisticated preparation, there would be this powerful synergistic effect, a biochemical version of the whole being greater than the sum of the parts?

If you ask most scientists, they will tell you it must have been trial
and error. And like so many scientific explanations for phenomena that
we don't understand, this is exposed as a kind of rude euphemism for the
fact that we have no idea how the Indians knew to combine these plants.
If you run statistical models you quickly realize that this (trial and error)
is simply not what happened. If you ask the Indians themselves, they
have, from their point of view, quite mundane explanations. They tell
you the plants told them. And, of course, we dismiss this because we're
uncomfortable with metaphor.

If you pursue it with them you'll find that peoples such as the Siona
Secoya of the Putamayo area of Ecuador and adjacent Colombia have
seventeen varieties of ayahuasca, which they consistently distinguish in
the forest, all of which are morphologically, to our eye, taxonomically,
the same species. When you ask them how they distinguish those, they'll
answer: "I thought you knew something about plants. Of course, you
take each plant on the night of the full moon and each species sings to
you in a different key."

Now, that's not something that's going to get you a Ph.D. at Harvard, but it's a lot more interesting than collecting and counting stamens.
It does give you an idea that in journeying into these distant realms, we
journey not simply to find new valuable species, which is sort of the
cliché of the purpose of ethnobotany. We journey in part in order to

return with new visions of life itself, to understand that the world which we live in does not exist in some absolute sense but is just one model of reality, the consequence of one particular set of adaptive choices that our lineage made, albeit successfully, many generations ago. If there's one revelation of anthropology, it's that there are other ways of thinking, other ways of being, other ways of orienting yourself on the Earth. And given our sorry predicament, having access to other ways of understanding may be vital to our survival.

When you think about that and you accept the biological fact that all peoples have the same mental acuity, an interesting question emerges: "What happens when a people don't put that mental ingenuity into creating cities and electronic technologies, but rather put it into exploring the metaphysical realm, or the realm of plants?" The work I did in Haiti showed me, powerfully, that different cultural realities create very different human beings whose capacities are profoundly different. Landscape *can* create character. Lawrence Durrell once said you could depopulate France, resettle it with Tartars and find to your astonishment that, within a generation or so, the same national traits would re-emerge—the love of fine food and wine, the reflexive disdain for Americans—all of this would just sort of pop out of the ground of the soil of France.

But culture also informs landscape. We have a kind of a Thoreau-an or Rousseau-an idea of the relationship between native peoples and the Earth that implies the same kind of self-conscious contemplation and separateness from the Earth that Thoreau celebrated. Thoreau could only speak about the wild the way he did because he never saw anything remotely wild in his life. Indigenous people throughout the world are not nostalgic or sentimental. They have forged through time and ritual a traditional mystique of the Earth that's based not on an idea of being self-consciously close to it but on a far subtler intuition, the idea that the Earth itself is breathed into being by human consciousness.

You can see this in any number of ritual practices and, of course, what's important is not whether the metaphors they invoke are true in some factual sense but what they tell you about a people. For example, I lived for a long time with the Cogi in the Sierra Nevada de Santa Marta. Their society's ideal is to abstain from sex, eating, and sleeping and, as much as possible, chew sacred coca leaves and chant to the ancestors. To this day, on a blood-stained continent, the Cogi are ruled by a ritual priesthood. Training for that priesthood involves (ideally) the acolytes being taken away from their families in infancy; sequestered in a cave in a world of shadowy darkness for eighteen years, through two nine-year periods deliberately chosen to mimic the nine months of gestation in the natural mother's womb, now metaphorically the womb of the great mother.

During that entire time they're inculcated in the values of their society, values that maintain the proposition that only their prayers maintain the cosmic balance. And after eighteen years of this, they are to be taken out before dawn, and everything they've learned in the abstract is affirmed in stunning glory as they see their first sunrise at the age of nineteen. Now this is a rather astonishing example of ritual played out through time, through culture, through landscape, but it is far from unique.

I recently made a film for the National Geographic of an amazing ritual that happens outside of Cuzco, Peru in the highlands village of Chinchero, the site of the summer palace of the second of the great Incan rulers, and I took part in this grueling ritual. Once each year the community runs the perimeters of its land. Each hamlet seeks the fastest young boy, who dresses head to toe as a woman and in so doing, becomes a *huayllaca*, a pejorative term for the feminine essence. These *huayllacas*, carrying the banners of their community, lead all able-bodied men on a race around the perimeter of their mountainous lands, an arduous feat.

You begin at 11,500 feet, drop down 2,000 to 3,000 feet to the base of the sacred mountain, then climb 3,000 feet through a series of sacred points (itos) where prayers are made and coca is given to the Earth. Then you drop 3,000 feet, climb another 4,000 feet and so on for 25 kilometers, all the way around. By the end of the day you emerge in a trance, almost finishing the race less as a human being than as a spirit being who, through the power of the collective, has been able to get through this incredible ordeal. I personally got through it by chewing more coca leaves than have ever been chewed by a human being in the history of the planet.

That notion of moving through sacred landscape often plays out in the use of sacred plants as well, most notably in the famous cult of the cactus of the four winds, which involves the ingestion of San Pedro cactus, *Trichocereus pachanoi*. I spent some time in the valley of Huancabamba, working with the *maestros* and the *curanderos* there. Acolytes and patients come together from all over South America to these baroque, nocturnal rituals, where they ingest through the nostril about a liter of alcohol infused with either datura or its leaves, and then drink a concoction at midnight of the San Pedro cactus, and during the ensuing intoxication, a diagnosis is made. The next day, in order to carry out the prescribed healing, one must move through sacred geography on a pilgrimage to a series of isolated lakes called Las Huaringas, found higher up in the mountains around the periphery of which grow the medicinal plants that are alone believed to be efficacious healing agents.

Again we see the metaphor of moving through sacred geography and realigning with the spirit level by making some kind of sacrifice so you can be open to the pharmacological potential of the medicinal plant. The relationship between healing landscape and consciousness is very strong in these societies. A kid raised in the Andes to believe that a mountain is the domain of a spirit that will direct his or her destiny will be a profoundly different human being than an American kid raised to believe that a mountain is a pile of inert rock ready to be mined.

I was raised in the forests of British Columbia, believing that a forest exists to be cut. That made me very different from a Kwakiutl youth who was raised to believe that the forest was the abode of the Crooked Beak of Heaven and the cannibal spirits that dwell at the north end of the world.

We have to be very careful not to project our own cultural categories and assumptions in attempting to learn from indigenous peoples.

Misconceptions abound. For instance, many people in the West seem very confused about what shamanic medicine is all about. In our society we distinguish between the priest and the physician. The physician treats the body and, for some, the priest has domain over the soul. We often read that in the shamanic traditions the priest and the shaman become one, but that's an oversimplification. There are two very different levels of treatment. On the one hand these cultures treat many diseases symptomatically, much as we do. In place of medicinal drugs they use medicinal herbs, many of which are pharmacologically active. But my experience has been that that's often the realm of women who are responsible for treating their families for all of the basic afflictions that affect their society.

The shaman's notion of healing is that diseases he or she is called upon to address are always seen to be in some kind of metaphysical realm. So shamanic healing requires that the shaman elevate his or her spirit to soar away on the wings of trance, to get into those metaphysical realms. It's a bit of a fantasy that the shaman is invariably a great repository of practical herbal medicine. Often it's the women who actually know more about that type of herbal medicine.

One crucial thing that distinguishes our lineage from hunter-gatherer cultures is, of course, the fact that we succumbed to the cult of the seed and that ten thousand years ago the poetry of the shaman, as Joseph Campbell said, was displaced by the prose of the organized priesthood, and we began to settle down, create surplus, hierarchy and all the institutions of a settled society. I'm mostly interested in the time before that and have spent as much time as I could with nomads: Inuit in the Arctic, pastoral nomads in Kenya and Tibet, or rainforest nomads such as the Penan in Borneo. All of these people are profoundly different than we are because they have not succumbed to the cult of the seed. They are different in a wide range of often startling ways.

What's it like, for example, to live with almost no material possessions? In nomadic societies there's no incentive to have many material possessions because everything has to be carried on your back. Their measure of wealth, quite literally, is the strength of the social relationships amongst people. That makes for a profoundly different sense of community. If they do not get along in a cluster of perhaps eight or ten people, and a group fractures, it means they'll have a 50 percent less chance of securing food.

Cultures take different adaptive paths, and it makes for profoundly different human beings. In nomadic society, sharing becomes an involuntary reflex, because you never know who's going to be the next person to secure the food. We walk past a homeless person in the streets, and we think it's an inevitable, perhaps tragic aspect of our economic system. But for many nomadic people the existence of a poor man is believed to shame everyone in the group. I've been with Penan and Gabra people in the streets of Europe or North America, walking past a homeless person, and they just can't believe it. It just seems incomprehensibly bizarre to them that a place so wealthy could tolerate such inequities.

Another major difference is that hunting is central to most of these cultures, and in those cultures that use visionary plants, their use is often closely associated or somehow linked with hunting. The Huichol, for example, explicitly refer to the seeking of peyote as a hunt. They call it "the tracks of the sacred deer." Shamanism was the first world religion, and religion emerges at that moment when people try to come to terms with the inexorable separation between life and death. How people resolve that separation determines their mystical world view, and hunting societies always had to come to terms with the terrible fact that to live they had to kill the thing they loved most, the animals upon which they depended.

In most hunting societies, whether in the rainforest or the Arctic, myths arose about the covenant that existed between predator and prey as a way of rationalizing that terrible fact. We are far more separated from what is killed in our name and lack myths to reconcile ourselves to our society's predation.

Most startling perhaps, when you study and live with, say, the Kalahari bushmen or tribal aboriginal people in Australia, you realize that these peoples never embraced a cult of progress or the notion of cultural change as we understand it. In the Australian dreamtime for example, change is really not a possibility because the world is understood to exist in two parallel universes, one of which is constantly being brought into being by consciousness, by individuals walking the song-lines: mimicking through gesture and song the songs of creation that the ancestors breathed into being at the beginning of time. When you believe so firmly that you are actually making the phenomenological world as you're walking, breathing and living, how can you have a cult of progress? There's nothing to improve upon yet because you're still making it.

The over-arching point I'm making is that those in our culture who are interested in sacred plants tend to focus too much on the plants and not enough on the fact that rituals of all kinds and the use of these types of plants specifically are always rooted in culture, and rooted in a more general desire to change consciousness, which can be done in myriad ways.

As fascinating as they are, we shouldn't put too much on the shoulders of these sacred plants. We need to focus more on the wonder of human adaptation to a remarkable planet, and sacred plant use is one aspect of that. Because of our own obsessions with psychotropic plants we sometimes put more weight on them than the indigenous people themselves do. And practices vary tremendously from culture to culture. With the Mazatec in Mexico, for example women are often *curanderas* who use psilocybin mushrooms in their work. But in the Amazon, most of the *curanderos* are men, and because they have this very complex and sophisticated alchemy with the various ingredients of ayahuasca, it's drawn a great deal of attention.

In addition to this, most anthropologists were men who ignored something that is, frankly, just as interesting, and that is the way that women in the Amazon make food. This ignorance prevailed until a well-known anthropologist, Christine Hugh Jones, lived with the Barasana in the 1970s and really paid serious attention to how the women prepared manioc, the staple food. Manioc is a plant that contains toxins and has to be prepared very carefully. It's dug up from the fields, rasped, and made into a sort of a paste in a very elaborate process. The water is leached through and then eventually put into something called the *tipiti,* a kind of unusual woven structure that you can squeeze very, very tight to get rid of all the liquid and with it all the toxins.

It turns out that a complex ritualized mythology goes along with the preparation of manioc, between the anaconda and the *tipiti,* a whole series of stories and songs as elaborate as the shamans' repertoire vis-à-vis ayahuasca. Yet it's simply about food. Because of our own obsessions, the ayahuasca preparation and complex may seem more interesting or important than something which we might reflexively dismiss as just the making of food, but the people themselves don't draw that distinction, and the realm of the woman and of food is just as vital an activity and as deeply imbued with ritual and myth in many of these cultures as that of the shaman.

Furthermore, taking sacred plant use out of its cultural context can be a risky business. There's a romantic myth that a shaman is always a benign fatherly or motherly figure, but I've never met a shaman who wasn't somewhat crazy. That's their job. They're the ones willing to go into the waters that most of us don't even want to know exist. Shamans are often marginal to their communities. Most Indian people, like people everywhere, are just trying to get by, trying to deal with the challenges of the day-to-day life.

Shamans walk that fine line between enlightenment and psychosis, but they do have a society willing to welcome them back that recognizes and validates the esoteric knowledge they seek. But I've had experiences with close friends of mine from our culture who tried to live like shamans, taking a lot of psychotropic substances months at a time alone in the wilderness, and it was a recipe for madness, because there was no cultural matrix to bring them back to, or to bring their visions back to.

So cultural context is tremendously important in the use of ritual in general and sacred plants in particular, and there are real risks in thoughtlessly dabbling with them. But that doesn't mean these plants can't be of great use and of interest to us, and haven't been extremely important in our own culture.

For some of us who have, by circumstances of birth and history, been brought into a particular worldview that sees itself as a paragon of human potential, they can sometimes serve as an antidote to our cultural hubris. After all, half our marriages end in divorce, only 6 percent of us have elders living with us, the average eighteen-year-old American kid has spent two years watching television. We embrace obscene slogans like "24/7," implying a total obligation to work at the expense of family, and we seem to have the propensity to rip down ancient forests, tear holes in the heavens, and attempt to change the biochemistry of the planet.

So those of us who embrace these plants are often seeking another path, another way of seeing the world. I'm one person who is very happy to say that I inhaled. When we look at the narratives about the sea change of our society throughout the Western world in the last forty years—issues such as the new attitudes towards minorities and other cultures, the roles and rights of women, the environment—one key factor in that change of outlook is consistently expunged from the record: the fact that millions of us laid prostrate before the gates of awe beneath the power of one of these plants. I think one hundred and fifty years from now that will be recognized as one of the catalytic factors that helped shift the history of our civilization, hopefully in more benign directions.

In that light, the government's war on drugs is one of the greatest examples of what the historian Barbara Tuchman described as national folly: a nation in full possession of the facts nevertheless persisting in policies against its own best interests. One startling statistic that is constantly left out of the debates about drug use is the fact that by all accounts, some ninety or ninety-five million Americans have tried "illegal" drugs. Roughly five million people regularly ("regularly" being defined as once or more a week) consume marijuana, and maybe 500,000 regularly take some psychoactive substance. Roughly ninety million Americans have been exposed to these substances, had them sweep over their imaginations, probably drew—for the most part—something good from them, and then moved on with their lives. And that, in fact, in an anthropological sense, is exactly what happens when drugs come into a culture for the first time.

Drugs are usually ignored by historians, but they have, on occasion, played important roles at key moments. I used to say, half-jokingly, that the French Revolution was caused by caffeine. And it sounds silly until you remember that until the eighteenth century, you couldn't drink the water in any European city if you didn't want to get cholera or some other pathogen, so all beverages were fermented. The whole continent was mildly besotted in alcohol the whole time.

Then suddenly within a thirty or forty-year period three central nervous system stimulants came in: tea from China, chocolate from Guatemala, coffee from Brazil (and before that from Abyssinia). Because these had to be made with boiled water, that took care of the cholera and the various pathogens, and because they were precious commodities they were dispensed in houses known as coffee houses or chocolate houses, which became centers of intrigue. Instead of having your beer, you were wired on caffeine and you couldn't shut up, and perhaps you suddenly noticed that Versailles was a little bit bigger than your place. The first seeds of the mob that stormed the Bastille marched out of a café after drinking coffee.

If we legalized drugs tomorrow consumption would barely increase, in my opinion. I've never met anyone whose decision to use or not use an illicit substance has ever had anything to do with whether the substance is legal or not. Vis-à-vis drug use, the consequences of criminalization are well-known. They range from a corrupt judiciary to seizure laws that allow the authorities to take my house if a houseguest of mine has an illicit substance; the creation of a prison-building industry with an enormous lobby; the devastation of our inner cities and the incarceration of so many of our youth; the assault on indigenous peasants throughout Latin America; and the destruction of one of its oldest democracies—Colombia.

Andean people have used coca leaves as a vital sacred plant for four thousand years with no evidence of toxicity, let alone addiction, and the price of those leaves has now gone up so dramatically that in many cases they're not available for ritual use. And in other cases, for example in the northwest Amazon in the areas controlled by drug cartels, indigenous peoples of the Tukano, Barasana, and other groups have been brutally forced into virtual slavery, a system of debt peonage as labor for the processing of the coca paste.

I had an encounter with an agent from the Drug Enforcement Agency (DEA) that was revelatory; it really opened my eyes to the essence of the dynamics of the drug war. A very close friend of Dr. Andrew Weil's and mine, Tim Plowman, the most brilliant ethnobotanist of our generation, was the world's authority on coca. I had been very fortunate to serve as his field apprentice of sorts in South America for some time. Tim had done some research on coca, funded by the United States Department of Agriculture (USDA) and the research had shown that coca was a very nutritious plant with more calcium than any plant ever studied by the USDA.

This made it very useful for the traditional Andean diet that lacked a dairy product. In the early years it was even suggested that coca might have enzymes that enhance the body's ability to digest carbohydrates at high elevations. Clearly this was a benign medicinal plant (in its leaf form; obviously when it's concentrated in powder and paste, it's far more problematic)—the antithesis of the image that it has as a pernicious drug (cocaine).

In the late 1970s, because of this research, Tim was contacted about a job opening at the USDA near D.C. He came to Washington, where I live, and visited me, but he found out the job was really for the DEA, so he said to me: "I want you to apply for this job, so we can see what they're up to, but if you take it, I'll kill you." I went out to Beltsville, Maryland, to the USDA, and I went into this office and could barely see through the cigarette smoke. Then I noticed that the walls were covered with psychedelic posters from San Francisco in the 1960s and seized hookahs (pipes) and paraphernalia. I was looking at this DEA guy, and he seemed really familiar. I used to have a farm near Medellin (Colombia), just when the cartel was starting. At that point a lot of the cartel guys were sort of sleazy, Colombian hippies. This DEA guy had an open collar with a hairy chest, gold chains, a lot of gold rings, and one of those wristwatches with nuggets on it.

It turned out what he wanted me to do, after all this elegant research that Timothy had done, was go back to the Huayaga Valley (and by this point, the Huayaga Valley was aflame with the Sendero Luminoso guerrillas), go into the coca fields, find the biological entities that seemed to predate on the leaves and bring them back alive so the DEA could manipulate them and make them even more predatory and release them in the fields as biological controls of coca.

As I listened to this kind of Dr. Strangelove scenario, I suddenly realized I'd never actually seen this particular guy before, but I had met him a thousand times in Medellin, right up the road. He was exactly like the guys in the cartel. This guy's energy, his consciousness, was exactly the same as Pablo Escobar's. That's why this absurd war goes on. They love each other. They don't want this to end. None of these drug warriors have the slightest interest in "winning" the war on drugs.

This presentation took place at the Bioneers Conference in 2001.

PART TWO

Psychedelics, Science, and Ways of Knowing

4

Psychedelic Empowerment and the Environmental Crisis: Re-awakening Our Connection to the Gaian Mind

Terence McKenna

It's always tricky to talk about the environment because it is our mother, the matrix that sustains us, but all our discussions now gravitate towards the crisis that haunts it. Can biological diversity and human diversity somehow make peace with each other so that the glories of these domains of nature, and they are both domains of nature, are not compromised?

The planet is being challenged in so many ways. I was in England recently and the newspapers were full of stories about fissionable material and plutonium triggers for tactical nuclear weapons that were being peddled in the red-light district of Frankfurt. The health of the Earth and the health of our children can be held hostage by the least noble among us. In a sense, we're in a dark age. Starvation, the displacement of populations, the collapse of central authority—all of these things, terrible in themselves, also exact a terrible toll on the environment. The world is growing more complex every day, every hour, every minute. The number of inventions, the number of connections among people, the number of people, all these things are increasing, compounding our creative potential and compounding our problems in a mad race between what seem to be incredibly positive forces and incredibly negative forces. And in this kind of situation, of course, each moral person is challenged by the old Tolstoyan question, "What is to be done?"

We each have very limited resources, limited strength, limited funds. Where shall we put our shoulder to the wheel to do the most good? This is a very complicated question, because it's not simply about doing good, it's about being effective. In a sense, this issue was postponed for forty or fifty years while the male dominators played the capitalism/communism shuffle and the rest of us were the marks, conned by that particular bit of historical chicanery. Now, the veil has been rent. Marxism's status as a paper tiger has been exposed, and what we now hear of are trade agreements and free trade, basically a license to peddle crap everywhere.

Trade in physical items that have been manufactured out of the body of the Earth should be made as difficult and expensive as possible. The piling up of material goods, consumer object fetishism, is the force that is destroying this planet, not only where it is actively practiced—here, for example—but throughout the world where people yearn to practice it because the images being beamed down to them in movies and on TV tell them that without a three-car garage, without a sunken bathtub, without a sailboat, you are nothing. And the fact of the matter is, there isn't enough glass, metal, and fossil fuel in the near surface of this planet to deliver a middle-class lifestyle to all the people who have now been sold on that idea.

We were misled, blinded, and bamboozled by this great historical struggle between communism and capitalism because both distorted the environment, toxified vast amounts of land, and squandered the treasure of their peoples in unconscionable ways. But where do we go from here? I think we have to draw back for a moment from our various projects, our activism, our botanical gardens, our seed and biodiversity preservation initiatives, the programs that feed the hungry—all the work we each commit ourselves to—and try and get an overarching understanding of the larger process in which we're all embedded. If we do that, we'll see that the human world is caught between two great polarities: materialism and eco-consciousness.

I say materialism and not capitalism because I believe that capitalism can perhaps be reconstructed to serve human needs. Materialism cannot, because there is not enough matter around for us to fabricate it into our toys without drowning ourselves in the toxic byproducts of that effort. Materialism vs. eco-consciousness is not a very fair match because the materialists have all the money, all the guns, all the propaganda. What we have on our side is that, if they win, ultimately, most likely everybody dies or at least life on Earth is seriously degraded. How long will the battle go on before this perception makes its way into the headquarters of the other side? I don't know.

As things stand, we don't even have a theory, let alone a practice, of toxic disposal. What we have is toxic hiding. Shell games. Postponement. Tabling. We need a real method of toxic disposal. Until it comes, we continue to sequester radioactive and toxic materials throughout the world in the body of the Earth. Many of these toxic materials have half-lives of hundreds of thousands of years. In yesterday's newspaper it was reported that a Greenpeace vessel encountered a Russian frigate in the Sea of Japan. The frigate was dumping nuclear cores and nuclear coolant into the sea, where the Japanese population draws 30 percent of its fish catch. This, sadly, is typical, not an aberration.

Plants may have a role to play here. Solanaceaous plants such as jimson weed and datura and other species can sequester heavy metals, but this information has not been made the centerpiece of a crash program to clean up the land simply because human institutions just do not yet care enough. So far most government and corporate environmental restoration programs are largely cosmetic efforts designed to divert attention from some greater horror being perpetuated somewhere else. We have to be very sophisticated because our materialist enemies are incredibly sophisticated.

We need planet-friendly technologies. An obvious one is hydrogen. Hydrogen, when burned in an automobile or a boiler, combines with oxygen to produce water. It is a technology that is non-polluting, and by developing liquid natural gas storage and transportation technologies, we have proven that we can handle hydrogen. Hydrogen is highly explosive, but if you blow up a hydrogen loading or unloading depot, all you have is a very violent, localized explosion. You don't have planetary contamination of the water and the air.

Hydrogen is the direction in which we should be moving and the people who have studied it have designed many sustainable technological schemes for bringing it to us, but the problem is that those who are already peddling natural gas and petroleum byproducts and nuclear power are not really interested. Their loyalty is not to the species or to the planet, but the bottom line. In a genuinely civilized society, one would think that putting the bottom line ahead of the entire civilization and planet would be a hanging offense, but not here.

Population control is another area in which we are in denial. A child born to a woman in San Francisco or Malibu will consume between eight hundred and one thousand times the natural resources of a child born to a woman in Bangladesh. And yet where do we preach birth control? That's right. Bangladesh. If every woman had one child, the population of this planet would fall by 50 percent in one generation. Without war, without famine, without displacement of populations, the population of the planet would fall by half in the lifetime of most people sitting here. We tend to assume the population problem is not resolvable. But, since children in the most industrialized countries use so many more resources than the children of the third world, if only 10 to 15 percent of the women right here in the high-tech, industrial democracies were to adopt a "one woman/one child" ideal, the release of pressure on world resources would be visible within ten to fifteen years.

We need to deconstruct product fetishism. We need to, somehow, through the media, establish the idea that a Zen-like spare-ness is the highest expression of social consciousness when it comes to interior decorating and the building of houses and so on. Let's deconstruct product fetishism so that the Kikuyu or the Yanomami become the paragons of behavior in terms of relating to material existence. We do not need to sell our souls to junk and then inherit a ruined planet. But the techniques of capitalism use advertising to spread discontent. They need to keep selling you new things you don't already have. Discontent is the obvious way to move you to make those purchases, so you are constantly being bombarded with images of your own inadequacy.

And at some point we will have to begin demilitarizing ourselves and become serious about democratic values and human rights. You cannot build eco-consciousness with an apathetic population ruled by oligarchs. We will have to wake up and make our governments accountable about preserving the environment, cleaning up toxic waste, and disposing of the weapon stockpiles that these characters feel they have to have.

I realize the laundry list I have just run through is not the sort of thing people are used to hearing from me, and it's easy to make demands. But how could we actually even begin to fulfill these many agendas at once? This is where we come to my bailiwick. It has been less than a hundred years since the great stabilizing and illuminating force in the lives of archaic peoples has come to the attention of Western civilization. The great stabilizing and illuminating force in the lives of archaic people is their vegetable connection to the Gaian Mind, their ability to experientially feel the planet—our mother—its needs, its tensions, where it is in pain. They do this to a great extent through the ingestion of psychoactive plants, just as they have been doing for fifty thousand to one hundred thousand years before the monotheist, agriculturist faction arose.

The greatest gift of the vegetable mind to the human order is the psychedelic experience because it allows the dissolution of boundaries, and it is going to be necessary to dissolve those boundaries in order to coordinate the metamorphosis of the human world. We have to have a vision. I don't mean a plan or an agenda. I mean a vision that comes from the unconscious—call it the Gaian Mind or the Great Spirit. It doesn't come out of committee meetings and the data gathered by statistical analysis.

The best leaders among us are no more than crisis managers attempting to manage us past an apocalypse that they are coming to believe is inevitable. That isn't good enough. We represent the cutting edge of novelty in the biological world. Our self-reflective consciousness is our great glory. It also opens for us a dimension of moral responsibility unknown to the rest of the denizens of nature. Part of our Promethean and godlike aspiration to the control of nature is the concomitant obligation to care for nature and to feel nature.

My message is one of psychedelic empowerment, of feeling. Ideas and logic are important, but first we have to *feel* our dilemma. If we could genuinely feel our dilemma and make other people feel our dilemma, we would move rapidly toward real solutions. But we are anesthetized. We are a world dying under anesthesia for lack of authentic experience, authentic connection with the living world out of which we came. We have to reach across barriers of culture and time and space to the essential humanness that unites all of us. And if we can do this, by whatever techniques are available, if we can create a sense of community, globally, then attending that sense of community will be a sense of caring and responsibility.

It cannot happen unless we change the way we think. We have the technologies, the money, the logistical ability to do almost anything in the human world, but we don't know how to change the way we think. It was all very fine when there were endless environments to despoil and vast herds of game. That curious amalgam of the animal and the angelic that is our humanness *could* exist in that kind of an environment, but no more. For the past thousand years the moral bankruptcy of Western civilization has become more and more apparent. The chickens are coming home to roost. We must build community and we must do it in a short time. If we had five hundred years to debate, I don't think I would be advocating psychedelic intervention. But we're sick. We're terminal. We don't know who we are. We don't know where we want to go.

Our lives are an experience of inadequacy and tension. We have lost our compass. Psychedelics are not a magical panacea, but they *can* lift the veil on the intention of the Gaian Mind. And we are but atoms in that Gaian Mind. If we do not follow its purpose, we *have* no purpose. Who do we think we are? Western science is six hundred years old. Human beings have been on this planet two million years, and life—1.4 billion years. There is an enormous wisdom in biology and we must become able to tap into that, articulate it, and activate it. We are the crowning achievement of the evolutionary process. Let's not betray it. Let's make it the ascent to angelic being that is, I am sure, the intention of the Gaian Mind and all the rest of the life with which we share this planet.

This presentation took place at the Bioneers Conference in 1993.

5

Plant Messengers: Science, Culture, and Visionary Plants

Dennis McKenna, Ph.D., Terence McKenna, and Wade Davis, Ph.D.

Dennis McKenna: My background was originally in ethnobotany and ethno pharmacology, but I've been working more recently in "straight" pharmacology. It was my interest in psychedelic plants, in what some call "plant messengers," that led me into this line of study. If we are to think of plants as "messengers," we have to realize that the language they speak is a molecular language. Plants, unlike animals, are chemical virtuosos. They can photosynthesize: take sunlight, carbon dioxide, and simple elements from mineral sources and, from those basic elements, form complex organic compounds (proteins, lipids, etc.—the chemicals that are universal in all organisms on Earth). Plants also go one step further and spin out a vast array of what used to be called "secondary" compounds with an enormous variety of structures, from alkaloids to terpenes to polypeptides.

Animals adapt to their environment mainly through behavior. They can move around. They can flee in case of danger or move toward something they're attracted to—a potential mate or a food source, for instance. Plants can't move, so they substitute biosynthesis, chemical ability, for behavior. They have elaborated an enormous array of chemical compounds to help them achieve their goals. Twenty years ago plant biochemists used to think these chemical compounds were just metabolic waste products with no real purpose, but within the last fifteen years that view has been changed, radically. It's now understood that these chemicals are often messenger molecules that plants release into their environment to influence their interactions with other organisms, be they other plants, or bacteria and fungi in the soil, or herbivorous animals or human beings that feed on them and use them in all sorts of ways.

Plants modulate and orchestrate their relationships to all these other organisms through these chemical messengers, called allelo-chemicals. A good example is an oak tree. If you look underneath an oak tree you'll find that the ground is bare. Nothing grows underneath the oak tree because it produces a flavonoid that is a very powerful plant growth inhibitor. In order to mark out its territory and prevent competitor species from impinging on its territory, the oak tree produces this flavonoid in its leaves. As rain falls, this chemical leaches out of the leaves and into the ground. It prevents anything from growing in the vicinity of the tree for a space of forty or fifty feet around it. That's one simple example of a plant's chemical interaction with other organisms.

What we see in plant chemistry—this variety of chemical interactions plants have with other organisms—is the result of four and a half billion years of evolution. In the case of the angiosperms, the flowering plants, this evolution has only encompassed the last sixty-five million years or so, but that's still a long time. So the chemical sophistication of plants is the result of millions, if not billions, of years of biochemical warfare during which the plants evolved chemicals designed to attract pollinators, for instance, or keep away potential predators. And the organisms these strategies are directed at also evolved adaptations for dealing with the plants.

Butterflies, for example, (and certain other insects) have evolved enzymes that are capable of detoxifying plant toxins. In some instances butterflies actually consume the toxic plants, take the toxins into their bodies and sequester them, so that the plant toxin in their bodies renders them unpalatable to predator species. This is a complex symbiosis between the plant—which is originally the source of the toxin—the butterfly, and the predator that preys on the butterfly.

Again and again in nature you see these complex chemical interactions resulting from evolution's dynamic creativity, as organisms keep trying to adapt and devise strategies by trial and error to use and manipulate each other. Plants, facing a successful chemical defense by an insect, in turn elaborate over time even more complex toxins, perhaps targeted at other cellular targets, other enzymes, other receptor systems, and it goes on and on. Eventually you see the complex web of chemical interactions that characterize these floristic ecologies.

But what's curious when it comes to the psychedelic plants is that these plants evolved neurotransmitter-type molecules that have remarkable properties to affect human brains long before there were nervous systems for them to interact with, long before there were neurotransmitters or brains. They must have originally developed these chemicals for some other reasons, to influence something in their environment, and, much later, when brains and receptor systems evolved, it turned out these plant chemicals had extraordinary capacities to be internal signaling mechanisms within brains. Instead of mediating signals between different organisms in the environment, they somehow became adapted to mediating signals between sets of receptor systems in human brains.

Terence McKenna: Life is the horseman; matter is the horse. In other words, life rides the horse of matter, but life is not matter. Life is information. It's that concept—information—that is the bridge necessary when talking about plants as messengers. However we may attempt to de-condition ourselves as inheritors of Linnaean taxonomy, we inevitably tend to think of the biological world in terms of species, which, in a given historical moment, are static.* We know they transform on evolutionary scales of time, but as we encounter them, we think of them as static.

Yet if one had an extraterrestrial point of view and had not had the benefit of Linnaean taxonomy, evolution on this planet would look like a gene-swarm that, somehow, from matter, produced self-replicating polymers with a capacity for encoding experience in the form of accumulated efficacy. It might be more apparent to you that information was somehow kneading matter to its purposes.

Nature is mediated by pheromones. Nature doesn't speak in what we think of as language, but nature has sophisticated channels and capacities for communication. Nature does it chemically. The immune system seems to be a kind of halfway station between the advanced mammalian brain and much more primitive forms of organization because the immune system is a chemical system, but it learns. It remembers, and it recognizes the things it has stored in its memory. In an advanced mammalian brain this chemical system has become a chemo-electrical system that processes identifications and recognitions much faster and more efficiently.

And if one is sufficiently sensitive to nature and one obtains heightened perception by dissolving the boundaries of cultural conditioning, then one can become aware of this ocean of information which is moving back and forth chemically, electrically, behaviorally, in us and throughout all of nature. But this flow of chemical information can only become apprehensible if we drop all of our theoretical–construct building and listen.

Listening to nature is what shamanism is about. The planet yearns to communicate, and all nature is in fact language. We are somewhat anes-

---

*Linnaean taxonomy, first developed by Carolus Linnaeus in the eighteenth century, is a system of classification widely used in the biological sciences. It classifies living things into a hierarchy, starting with "domains" or "kingdoms," then divides into "phyla" (singular: "phylum") for animals; the term "divisions" is used for plants. Phyla are divided into "classes" and they, in turn, into "orders," "families," "genera" (singular: "genus"), and "species" (singular: "species"). Groups of organisms in any of these ranks are called "taxa" (singular: "taxon"), or "phyla," or "taxonomic groups."

thetized to this by our very introspective cultural style. Our whole focus of attention is inward, and so the natural world has fallen silent for most of us. Jean Paul Sartre said: "Nature is mute." That, sadly, captures perfectly modernity's relationship to nature, but still—if that isn't the lamest statement made by a twentieth-century philosopher, I don't know what is.

Dennis McKenna: If you juxtapose that statement of Sartre's with what the *ayahuasqueros* tell us, we can *really* understand the deep disconnect that presently exists between the way the modern Western mind views nature and how indigenous peoples do. If you ask an *ayahuasquero* how he learned about these plants and their medicinal value—how to use them and what they're good for—he'll say: "I learned about them in visions." They view the psychoactive plants as teachers, and *ayahuasca* as a master-teacher. The conventional wisdom among *ayahuasqueros* is that if you want to learn the properties of a new plant—what its medicinal or spiritual qualities are—you take it with *ayahuasca,* and *ayahuasca* will teach you about the plant in visions.

This way of knowing is related to the pheromonally mediated information exchange between plants and other organisms that takes place at a subliminal level. Fifteen years ago I was researching psychoactive virola snuffs with a shaman.* We collected a dozen different kinds and put them into a bag. Every time we'd collect one, he would strip the bark off the plant, look at it, smell it, taste it, examine its properties very closely, and then pronounce a judgment about that particular snuff. He would say that one was strong; one wasn't; one would be good to make an orally active paste from, and so on and so forth.

I noted all of the shaman's comments in a collection notebook. We got this stuff back into the lab where we have gas chromatographs and spectrometers, and I began to examine the samples. The shaman had said that one snuff in particular was strong. Therefore, clinical analysis should

---

*Virola is a genus of trees in the Myristicaceae (nutmeg) family, whose resins are widely used by groups in wet tropical forests of South America as hallucinogenic snuffs.

reveal a very high amount of tryptamine in it. The shaman had indicated that another snuff was not that strong. Thus there should either be no alkaloids in it or low levels of alkaloids.

What do you think the correlation was? There was a 100 percent correlation between what he said and my gas chromatograph's readings. The shaman didn't have fancy instruments. He used his highly refined senses to tune into the plant's language—its use of smell, taste, appearance, pigments, and pheromones—to trigger receptors in the nervous, olfactory, and visual systems of other species.

Wade Davis: I observed similar phenomena when I was working in Haiti. Joseph Campbell was once asked to name one place where people actually lived their religious beliefs, and without pausing he said "Haiti." People in the voodoo societies there walk in and out of their spirit realm with an ease and impunity that's astonishing. You become very much aware of the fact that this construct we call reality is just one model of reality, and I find that to be a very hopeful realization.

For me, one of the most intense pleasures of travel has been to be able to spend time with shamans who think they can go beyond the Milky Way, or with nomadic Penan hunters who can follow an animal call through the night. All of these lessons about the range of the world's cultural palette have taught me that this worldview we've inherited in our own culture is just one model of reality, and that there are other ways of thinking and interacting with the Earth.

It was in Haiti that I became absolutely convinced that indigenous people who have not bought in to a particular kind of positivist tradition but, by contrast, are open in their essence to the whole spectrum of life as they see it, are probably capable of intuiting things from the natural world that are simply closed to us. This is not because indigenous peoples are better or worse than we are, and not because they've got some special metaphysics. It's because all of these sensitivities are responses to specific cultural choices. Our sailors used to be able to see Venus during the daytime. That's something we can't do anymore, not because we can't do it physically, but because, culturally, it's no longer an attribute that we value.

I learned from my experiences in Haiti to really believe that people who are not brought up in a Cartesian world are capable of almost anything. Certainly you see that in the power of spirit possession. You can be sitting next to someone one moment, and suddenly she becomes a god from the voodoo pantheon. The spirit has displaced the soul of the living and that person has become a "divine horseman," mounted by a spirit. When you see that, you get a kind of visceral sense of the potential of the human imagination and human spirit unfettered by the constraints of Cartesian logic.

Terrence McKenna: Science told us that it was giving us objective data about a real universe that was not culturally conditioned. After all, the orbit of Jupiter is the orbit of Jupiter no matter what religion one practices or what culture one is from. It is not culturally conditioned, but quantum physics' awareness that the observer is as much a part of the system as the thing being observed undermines that type of certainty.

Can we really be sure science is not somehow ultimately as subjective as voodoo or astrology or sortilege as practiced by the Maya or the Yoruba? What we thought was the search for absolute, incontrovertible truth might turn out to be nothing more than the pursuit of a cultural illusion of truth. These people who stayed in the rainforest never fell for such epistemological naiveté.

A friend of mine who works with the Guarani told me they say: "We don't understand anything. We have no explanation. We have no explanation for anything."* It sounds to me as though these may be the most

---

*The Guarani (pronounced "Waraní") are one of the most important tribal groups of South America. Their former home territory was chiefly between the Uruguay and lower Paraguay Rivers in what is now Paraguay and parts of Argentina. They belong to the Tupí-Guaraní group, which extends almost continuously from the Paraná to the Amazon, including most of eastern Brazil, with outlying branches as far west as the slopes of the Andes. The Tupí-Guaraní dialect is based upon the *lingoa geral,* or the Indian trade language of the Amazon region.

sophisticated people on the planet, perhaps the *only* people on the planet who have freed themselves from the illusion of the cultural truth. Why don't we all just admit that we don't have a clue? Then these shamanic phenomena and these mysteries of other cultures become considerably less mysterious and greatly more accessible.

We're no longer looking at these things in a museum diorama; we can go to the center of the galaxy as the *ayahuasquero* does. We too can cure through the power of attention and concentration. This is what this abandonment of science holds out for us. A lot of constipation and false reassurance is put aside, but an enormous dimension of freedom and opportunity opens in front of us. I think you have to take your experience as your experience and not think that a shaman is necessarily seeing more deeply into reality than you are. You're never going to be an Amazonian shaman. One should make one's peace with one's own circumstance in the universe. We are all new here.

A shaman said to me once, "Don't think that it's easier for us just because we don't wear clothes." It isn't easy for anybody. All cultural structures pose problems in a quest for knowledge. A shaman is a person that a culture has agreed can go outside the culture. That's how the shaman does his magic. Everybody else is watching the TV; he's back behind it, putting in tubes, taking out tubes, checking out circuits and looking at the whole game.

Dennis McKenna: I'd like to defend science a bit. It's true that the goals and practice of science have often become perverted in our culture and misdirected. But, if science is practiced in its truest form, it is a tool for discovery, for learning about our world. It is a type of shamanic activity, and the kind of thing that the shaman does: pulling out and replacing the tubes of reality to discover what happens if you tweak one this way or that way. The shaman's activity is also a form of scientific endeavor in the true sense. I think that what science has largely lost sight of in our desire to establish scientific or technological fixes for all our problems is that the essence of the scientific stance is suspended judgment. If we really think we know what's going on a priori, then we've departed from the true scientific stance, and in addition to that, we're just deluding ourselves.

But I think that if you abandon science completely, then you tend to get gullibility. Science gives you the tools with which to ask critical questions. If you ask the question in the right way, you get feedback from the environment. You're supposed to get information from the environment that tells you if you're asking the right questions or if you're framing them in the right way. If you are, you can ask more intelligent questions, which gives you even more information and you can further the understanding of whatever phenomenon you're trying to investigate. Science is an accumulation of information that allows you to frame more questions, but you should never come to the point at which you think you have it all figured out.

There can always be a new piece of evidence that totally blows your theory apart and you're back to square one. In some ways we've lost sight of the true goal of science: to probe the unknown, to use it as a tool to understand the world. Science is not necessarily more valid than other sets of tools to understand the world, such as shamanism, but perhaps we can make these different ways of looking at the world work together to give us a more complete picture. But I certainly don't think you can dismiss science completely.

Wade Davis: I think it's important to see that modern science, at its origins, was extremely liberating. When rebel thinkers in the Renaissance realized it might be possible to no longer be bound to the tyranny of the ideology of the Church, that was a great period of liberation in our history. The only problem is that we eventually threw out the baby with the bathwater because we then went on to dismiss all the possibilities of myth, magic and metaphor. And one of the reasons we now do what we do to the Earth is because we have lost all sense of its magical resonance.

We can't throw out science, but we have to recapture some of our capacity for myth and magic. A young Canadian kid is brought up to believe that a mountain is a pile of rock; a young Quechua is brought up to believe it's a spirit. Is that mountain a spirit or is that mountain a pile of rock? Ultimately the question is irrelevant; the key issue is how that belief will affect the life of that individual. Clearly, a young boy or girl brought up to believe that a mountain is a spirit will have a very different relationship to the landscape than one brought up to believe it's a rock.

Terence McKenna: Where I part company with science is at that moment when Francis Bacon, who was the great theoretician of modern science, wrote: "Nature is a goddess that we may lead to the rack . . . and there tease, torture, and torment from her her secrets." We're way off base at this point because we've lost the sense that we are a part of nature, that it is our mother and that you don't lead your mother to the rack in order to torture her and pry loose her secrets. I would argue that by that time Francis Bacon wrote those words, it's no accident that science was involved in the perfection of gunpowder, the trajectories of ballistic siege machines, the production of lenses for spyglasses for use in battle, and so on.

Dennis McKenna: I have been privileged in the last couple of years to work with a syncretic religious group in Brazil that uses *ayahuasca* (or, as they call it, "hoasca") in a ceremonial ritual context. It is both the central ceremony and the central mystery of their religion. Through Botanical Dimensions, we have received funding to do a biomedical study of hoasca as it's used among these people. This relates to everything from psychological profiling of long-term users, structured psychiatric interviews, and some biochemical measurements, specifically of the metabolism of the *ayahuasca* alkaloids in the body and of hoasca's potential long-term effects on serotonin receptor-systems. (This is the "Hoasca Project" that is referred to at various places in this book.)

In June and July of 1993, we completed the field phase of the project. Dr. Charles Grob and I, and several other Brazilian, Finnish, and American psychiatrists spent five weeks in Brazil working on this project, with the help of the members of this group. We've completed the field phase and the psychological portion of the work, and have now brought the blood and plasma samples back, and at this point we're waiting for our colleagues to complete the biochemistry work. We hope to have all this completed by the end of the year and will be submitting papers to the usual gamut of journals.

When I was in Brazil, several times the question came up: "What do you hope to learn by doing this study? You're never going to understand *ayahuasca* by making these biochemical measurements and by doing all this statistical analysis. That will never explain the central mystery." My answer was always that I didn't ever expect the study to explain the central mystery or what is truly magical about *ayahuasca*. Our goals for this study were actually very modest. Since there was no data on the biomedical aspects of *ayahuasca,* our project was an exercise in data gathering. We're trying to use the tools of science to carry us as far as we can go in understanding the drug's psychological phenomenology and its biophysical actions on the body. Never did we think this was going to throw open the drug's deepest secrets. I'm completely aware that after all is said and done, there will still be a profound mystery that science is unable to address. But I'm a scientist and I think it is worth investigating those aspects of the mystery that can be studied through measurement and careful statistical analysis. It's part of an overall effort of understanding.

People ask me sometimes how I can reconcile my scientific worldview with some of the mysteries I've encountered in the realm of psychedelics, such as encounters with what seem to be other forms of non-corporeal intelligence. How can a materialist, reductionist view explain experiences of what seem to be conscious forces ensouling or organizing nature, for example. My answer is that matter by its very nature is a self-organizing phenomenon and that organization equals information. It seems to be an intrinsic property of matter that, given the right circumstances, it tries to organize itself to whatever level it can, and if it happens to be in an ecology or environment where it can flourish out into biological forms of organization, then so much the better. It's not that there's a teleological force guiding evolution, in my view, it's just that information and organization are inherent properties of matter.

And consciousness is not an epiphenomenon of matter so much as an epiphenomenon of organization. A metaphor that appeals to me can be drawn from relativity theory in which gravitation is understood to be a property of mass. When you get enormous mass gathered together in one place, then you get what's called a singularity or a black hole where accumulations of mass cause the curvature of space-time. What's singular about the singularity is that you can't say anything about it. The conventional laws of physics don't apply within it. Anything could be going on in there. One has no idea.

The analogy to consciousness is that I have the feeling that consciousness is a property of high levels of organization. When you get a very high level of organization accumulated in a small area, such as the level you find within the human brain, for example (and it can be argued that the human brain is the most highly organized piece of matter that we know about), then you get this emergent phenomenon of consciousness. In other words, the human brain might be to organization what the black hole is to gravitation. Unpredictable, mysterious properties emerge out of such an amazing density of organization.

This discussion took place at the Bioneers Conference in 1994.

6

The World Spirit Awaits Its Portrait: True Tales from One of the Planet's Great Visionary Artists

Alex Grey

In 1982 in New York, I did a performance called *The Beast* in which I sat in a pool of black, tar-like liquid dressed in military gear. Behind me was a giant multi-limbed skeletal beast caught in a web hanging on a world map. As people entered the space they saw a hydrogen bomb blast projected on one wall to one side and a mural-sized painting entitled *Nuclear Crucifixion* on the other. I offered to stamp their hands with the number of the beast—666. Some people wanted me to stamp their foreheads.

My wife, Allyson, joined me in another performance piece called *Wasteland,* in which we represented the nuclear family. On each seat in the audience a pamphlet explained: "Mr. and Mrs. X were on their way to dinner when they were surprised by a nuclear blast. They arrived at a dinner table in Hell to feast on money." We sat at the table with a skeleton, drank "blood," and ate actual money. An alarm bell was going off in the heart of a skeleton and the sound of bomb blasts could be heard in the background. As the "clock halo" behind the skeleton's head approached nuclear midnight, Mrs. X got up and, being a gifted bulimic, vomited the money and blood onto a table. Then she jammed her fingers down my throat and I vomited too.

For a performance entitled *Human Race* I created a machine/vehicle with a long "bed" anchored into a cement floor on one side with steel rods. At the other end was a motor, a wheel and a clutch with a hand throttle to run the vehicle round and round. The audience sat about a foot away from the wheel as it whipped around. Once the machine started up, I lay down on the bed and the contraption started going round. I imagined the piece would end when the noxious fumes eventually drove out the audience, or it would just run out of gas. But the wheel started picking up more speed than I had anticipated, and it then occurred to me I hadn't put a brake on the thing. It got going faster and faster until it sheared the steel rods right off and careened toward the audience. Fortunately, I was able to fall off and stop it before it hit anyone. Everyone jumped up in wild applause, thinking that the ending was planned and perfect—the machine out of control.

We later did several performances that included the essence of different religions. In *Burnt Offering,* I came out dressed in a loincloth with blood poured over my body. I read about holy fire from the *Baghavad Gita,* the Bible, and the Koran; sacred books of three different spiritual traditions. After each reading I placed each book in a kind of totem urn made of animal and human skulls and set the books on fire. When the fire burned down, I mashed the ashes together and rubbed them on my body. Covered in blood and ashes, I sat before seven skulls with a skeleton behind me and lit a fuse that set them all on fire.

Another performance in 1983 was named *Prayer Wheel,* after the Tibetan device that devotees spin as a way of generating good will and furthering one's own spiritual evolution. This piece was about the life cycle and the individual as a combination of polarities, embodying male and female, birth and death. To embody this I tied a skeleton on my back while Allyson carried a baby doll. To represent the great value of the soul, we painted our bodies and hair gold. We were tied together and tethered to a giant eight-foot prayer wheel we created, which was illuminated from within. The inner light lit up the six-syllable national mantra of Tibet, "Oh hail the jewel in the lotus"—the jewel being our spiritual essence within our lotus-like lives.

Allyson and I have used diverse and universal symbols in performances —the yin/yang symbol, the cross, and the Star of David—to highlight polar opposites such as heaven and earth or matter and spirit, for instance. In the performance *Living Cross,* we lay together, surrounded with roses in the center of a giant cross made of five hundred apples outlined by votive candles. Above our bodies was a winged angel of death and transcendence, holding an eight-foot neon infinity symbol. Gregorian chants played in the background as we lay on our backs, naked and motionless for three hours.

For an outdoor performance at Lincoln Center, we used 5,500 apples to make a forty-foot effigy of a goddess. I did a hundred prostrations at the foot of the Goddess while Allyson, sitting in the heart center, nursed our daughter, Zena, for the last time. After we completed the piece, it remained on view for an hour or two. Then we boxed all the apples and donated them to homeless shelters.

Heart Net was a yearlong participatory installation we did at the American Visionary Art Museum. Hundreds of flowers created a heart shape with a stained glass eye in the center that was crying into a pool. A baby with the body of the world globe looked as if it had crawled out the pool. On the wall, a giant net hung over a sixty-foot map of the world. People were invited to write prayers or good wishes for the planet, for healing, and so on, and hang these messages on the net. A Buddha figure positioned at the top visually bridged the Earth to a transcendental realm. By the end of the year, thousands of prayers had accumulated on the heart net.

A six-hundred-pound bronze sculpture called *World Soul* took me two years to complete. It's a hybrid, divine mutant, multi-faced, self-copulating, hermaphroditic dwarf, perched on a world globe. It has a fish tail, claw-like paws grasping the Earth, and eagle wings. As a kind of shamanic amalgam of elements, the fish points to the subconscious, the paws to the Earthly realm, and the wings indicate the super conscious. I received this image after an exhausting day of teaching when I was simply standing on a subway platform.

When I first saw this vision, I thought it was a painting image, but then it opened up its wings and turned around and I could see it was a sculpture. The character points to its own heart center, which is a mirror, as if to say: "See yourself reflected in me; I am you." Its many-eyed head sees all around. Of its four faces, one has a wrathful aspect—a demon cradling new life, an infant in its arms. The infant, with its hands in a teaching *mudra*, stands as a symbol that implies: "Consider the generations to come before you ruin this gem of a world." Before welding the head to the body and sealing the bronze piece together, we encased precious items inside its heart: scrolls of prayers from many world religions and shamanic invocations, objects that had been blessed by different religious leaders, and a number of plant and other entheogenic substances and medicines from around the world. I was inspired by African sculptors who sealed magical elements inside of their figures to bring them life and empowerment.

High Times magazine asked me to do a poster for their "Cannabis Cup," an annual event in Amsterdam in which the best pot in the world is judged and awarded. My wife and I were invited to Amsterdam to be celebrity judges. It was a difficult job, but someone had to do it.

Obviously, entheogens have been central themes in my art and life. I did a painting portraying Adam and Eve as early humans eating from the tree of the knowledge, the tree of good and evil, with the forbidden fruit as entheogenic plants and mushrooms. The earliest religious book ever written—the *Rig Veda*—contains hundreds of references to Soma, a sacrament that would allow one to see God. During initiation into the Eleusinian Mysteries of ancient Greece, people drank an entheogenic potion called the *Kykeon* to experience a vision of divine reality.

Clearly the mystical foundations of both Eastern and Western civilizations contain references to psychedelic use. We know of the shamanic use of consciousness-altering plants from Central and South America and throughout the world. I hope that in the future our society will once again find a place to partake of the gifts and power of entheogens in our spiritual lives without risk of harassment and incarceration.

The *Sacred Mirrors* series, now my most well-known body of work, developed out of a performance called *Life Energy* that Allyson and I did in 1978. To explore life energy in various ways we did a series of exercises with the audience. As part of this process I created two charts of human figures. One was of the nervous system, a Western model of life energy, and the other was a more esoteric model with auras, chakras, and meridian points. We demarcated a zone in front of the charts suggesting that viewers stand in that zone, identify with the figure, and use it as a mirror for imagining the systems in their own bodies.

At the end of the *Life Energy* performance I executed a rat in order to show the passing of life energy. This didn't go over too well with the audience, and we thought we might have lost all our friends. As Allyson and I were walking home, a little despondent, she said: "You know, Alex, people really liked the charts. You should do a whole series based on them." That was the birth of the *Sacred Mirrors*. It evolved, and eventually I did an exhibition at the New Museum in New York. These life-size figures, each about six feet high, have frames that are five feet wide, ten and a half feet high, with stained glass in the crest. The twenty-one *Sacred Mirrors* depict the body, mind, and spirit of the individual.

The first *Sacred Mirror,* one called *Material World,* is comprised of dozens of mirrors sandblasted with the periodic table of the elements. In the center is a six-foot lead figure with biochemistry of the body hammered into it. This piece represents the first step on the journey, the material plane of elements and chemistry.

The elaborate frames of the *Sacred Mirrors* are covered with symbolic imagery. One side represents biological evolution and the other side technological evolution. A big bang is depicted at the bottom of the frame and evolutionary symbols ascend up the sides. Biological evolution is portrayed as a DNA chain rising up one side. Unicellular and multicellular life forms, on the rungs of evolution, graduate eventually to higher mammals. Technological evolution—from stone tools to the space shuttle and brain science—are all represented in a twisting chain of Ouroboric snakes. Depicted at the top of the frame are the polarities of male and female and a ring of symbols representing life and wisdom paths. A stained glass eye of God, cosmos, Spirit, in the top, expresses universal self-awareness and the evolution of consciousness.

I taught myself anatomy by painting the skeletons and bodily systems in the *Sacred Mirror* pieces. I wanted people to have an experience of the systems of their own bodies and their infinite complexity. The eyes in the paintings are open and staring, to offer a focal point for the viewer. The first paintings in the series look beneath the skin at the skeletal, nervous, lymphatic, viscera, and muscle systems. The next six *Sacred Mirrors* depict the largest organ of the body, the skin. Once there is skin, a specific gender and race is apparent. Portraits of male and female humans of various races represent the "mind" portion of the series. The mind constantly differentiates between self and other. We see the visible differences in individuals while seeing our self-reflection.

The next part of the series leaves the purely physical realm and takes the viewer to the layer of subtle energetic and esoteric anatomy. The *Psychic Energy System* shows chakras and acupuncture meridians and points, auras, and a pranic ocean of light surrounding the figure. The *Spiritual Energy System* portrays a breakdown of the boundaries between the Self and surroundings. The *Universal Mind Lattice* is a depiction of the total dissolution of the physical plane.

This image came from an LSD experience that my wife and I had in 1976. Allyson and I got together on my first LSD trip, and for us LSD has served as a kind of magical elixir. We sometimes lie in bed and take a mega-dose, put on blindfolds, and experience the complete meltdown of physical reality. On the special occasion that resulted in the painting *Universal Mind Lattice*, we became toroidal fountains of light interconnected in an omni-directional field that extended infinitely.

Every other being and thing in the cosmos appeared as one of these fountains of light, these cells of energy, and the energy that was going through all of us was the same—the energy was love. Everyone and everything was a part of this vast love circuit, yet each of us was a distinct node in this field. Staring endlessly into space, it was visually clear to me that everyone and everything was connected, while simultaneously realizing our uniqueness and essentiality. It was truly one of the most beautiful and strangest experiences ever.

After "floating suspended in the net" for a while, we took off our blindfolds. Allyson said, "I was in the most amazing place," and she started to draw. She drew exactly what I had seen. In the dark and without communicating we had shared identical visions of the same transpersonal reality. This blew our minds and totally changed our work. From then on, we determined to make art about this experience of sacred interconnectedness. The *Universal Mind Lattice* attempts to capture that vision.

The next painting points to the void-ness, clarity and awareness at the core of all mystical teachings, a space beyond depiction. *Void/Clear Light* portrays the elements—fire, water, earth, air—and the *Kalachakra,* the Tibetan Buddhist symbol of the wheel of time and the principle of the transmutation of the elements by the principle of emptiness. My painting of the deity *Avalokitesvara* shows the thousand-armed Buddha of active compassion. All the hands reach out like activists, with an eye of unobstructed vision in each palm, enabling a vision of what needs to be done to relieve all suffering beings.

By including a number of mystical traditions in the *Sacred Mirrors* series, I mean to go beyond dogma and seek common threads. The painting of *Sophia,* the Goddess of Wisdom, comes from my imagination, as there are few conventional portrayals of this deity. She is an archetype of the feminine aspect of the Godhead and is made of eyes, a symbol of awareness. Consciousness has no color, no weight, no form, so I use the symbol of multiple eyes to point to the concept of infinite consciousness. The final work in the *Sacred Mirror* series is called *Spiritual World*. This piece is, in fact, an actual mirror sandblasted with a sunburst lattice of radiance. Standing in front of it, the Sun becomes the head of the viewer. In the center of the sunburst is the word "God" etched into the glass, pushing the point that the *Sacred Mirrors* invite us to see ourselves, each other, and the world as reflections of the divine.

Besides the *Sacred Mirrors,* I've done numerous paintings that look below and beyond the skin at human physical and energetic systems. The series entitled *Progress of the Soul* depicts the spiritual and esoteric dimensions of the human path from conception to death. The painting *Kissing* shows an X-ray view of the physical systems of a male and female embracing. In the painting *Copulating* the couple, mortal flesh, genitalia, bones, muscles, and nervous systems entwine. Also revealed is the infinite element of consciousness, through the use of golden flames and bands of golden infinity symbols looping through the minds and hearts of the lovers-vortices that shoot out from the lovers alert waiting souls in another dimension of an opportunity for incarnation. This painting draws from the Tantric tradition the image of lovers connecting with each other not only through the skin but, in a sense, fusing together at the heart level and dissolving into each other.

The painting *Pregnancy* uses a lotus as a symbol of the soul. In the dynamic painting called *Birth* I attempt to capture the incredible channeling of almost explosive energy that goes through a mother during childbirth as well as the ultimate compassion inherent in giving birth. I use Tibetan seed syllables to express: "Here is a birthing Buddha." A painting called *Nursing* once again captures the physicality of the act as well as representing the subtle fields of interconnectedness between mother and child that forge invisible but powerful emotional and spiritual bonds. Our daughter Zena is the subject of several paintings at various stages of her life.

The *Dzogchen* teachings in Tibetan Buddhism state that our inherent Buddha nature is one of primordial perfection, but that it is, more or less, obscured. It's sometimes easier to see primordial perfection in children. In my paintings *Reading, Family,* and other works that include images of Zena, I intend to express family bonds, the passing on of culture and knowledge and the development of conscious awareness.

Certain paintings deal with archetypal figures and themes. *Painting* depicts a cosmic funnel of inspiration entering the artist's brain. A thanatopic phantasm behind the artist suggests that he better get to work, as life is short. A "peanut gallery" of demanding critics—including Van Gogh, Michelangelo, William Blake, Rembrandt, Frida Kahlo, and other renowned artists in the background—looks on critically. The painting *Vajra Brush* utilizes the image of the Tibetan *Vajra* scepter incorporated into the paintbrush, to represent a spiritual tool. There are allusions to prehistoric and other artworks in this painting as well, implying that our individual creative energy draws from a collective field of humanity's shared consciousness and traditions.

Art has been a way to integrate both the most difficult and the most uplifting material in my life. I don't feel complete unless my art can express a full emotional range. I experience darkness and the shadowy sides of life all too often. Those aspects are components of all great works of art, no matter how exalted, and that's why I point to them in my work as well. The paintings *Caring* and *Dying* depict the dying process in two stages, and consciousness is again symbolized by eyes—one of my recurring motifs.

A 1989 painting entitled *Gaia* contains realistic depictions of the planet's current environmental and political crises. It was brought to my attention after 9/11 that in that painting there are two airplanes flying over the twin towers. Three chummy conspirators stand on the destroyed side. One looks ominously like G. W. Bush with his arms around a "Dick" (a diseased phallus) and a terrorist.

Several major works I've painted deal with the human spiritual quest. *Holy Fire*, an eighteen-foot diptych, portrays a pilgrim on a mountain receiving divine grace in the form of a lightning bolt into his heart. His physical body then melts into a sun-like radiant fire. The painting describes the dislodging of the identity from the material body. A seven-paneled sequential altarpiece called *Nature of Mind* follows a pilgrim on life's path. It portrays episodes in that journey, from the discovery of sacred texts to the appearance of a teacher to an experience of enlightenment, concluding with his return to society to share his newfound wisdom.

Though this is not a fashionable view today, I believe that one of the artist's functions is to fearlessly probe behind appearances and illusions in a quest to experience clarity and universal truths and to then seek to communicate those experiences. I will end with this statement from my book *The Mission of Art,* which captures some of what I feel art can be:

"Art can be a form of worship and service. The incandescent core of an artist's soul, a glowing God's eye, infinitely aware of the beauty of creation, is interlocked with a network of souls, part of one vast group soul. The group soul of art beyond time comes into time by projecting symbols through the artist's imagination. God's radiant grace fills the heart and mind with these gifts of vision. The artist honors the vision gifts by weaving them into works of art and sharing them with the community. The community uses them as wings to soar to the same shining vistas and beyond."

> *Translucent wings team with eyes of flame*
> *on the mighty cherub of art.*
> *Arabesques of fractal cherub wings*
> *enfold and uplift the world.*
> *The loom of creation is anointed*
> *with fresh spirit and blood . . .*
> *Transfusions from living primordial traditions*
> *empower the artist,*
> *shaman, yogi, devotional prayer—*
> *all break through with the visionary cure*
> *and take the artist to the heights and depths*
> *needed to find the medicine of the moment,*
> *a new image of the infinite one,*
> *the God of creation*

manifesting effulgently, multi-dimensionally,
with the same empty fullness that Buddha knew
and the same compassionate healing that Jesus spread. Krishna
     plays his flute,
the Goddess dances,
and the whole tree of life vibrates with the power of love.
A mosaic and tile maker, inspired by Rumi,
finds infinite patterns of connectivity
in the garden of spiritual interplay
as the World Spirit awaits its portrait.

This presentation took place at the Bioneers Conference in 2003.

PART THREE

Sacred Plants and Human Cultures

7

The Garden and the Wild— Plants and Humans: Who's Domesticating Whom?

Michael Pollan and Wade Davis, Ph.D.

Michael Pollan: I'm interested in very ordinary plants—apples and tulips, cannabis, and potatoes. I think a different kind of power accrues to us if we can start to see things from the perspective of other species, and plants in particular. That's a key imaginative act toward re-conceiving our relationship to the rest of the world, overcoming this very parochial perspective we have about ourselves as the only subject in nature.

For most of history, gardens were grown for the power of their plants rather than their beauty. In ancient times, people all over the world grew or gathered sacred plants and fungi that had the power to inspire visions or take them on journeys to other worlds. Some of these people returned with a kind of spiritual knowledge that underwrites whole religions.

In the Middle Ages, apothecary gardens focused on species that healed and intoxicated and occasionally poisoned. Witches and sorcerers cultivated psychoactive plants. Their potions called for such things as datura, opium poppies, belladonna, hashish, fly-agaric mushrooms, and the skin of toads, which can contain DMT, a powerful hallucinogen. One use of some of these ingredients was to combine them in a hempseed oil-based "flying" ointment, that the witches would then administer vaginally using a special dildo. This was the "broomstick" by which witches were said to "travel." It's only been in modern times, after industrial civilization concluded, rather prematurely, that nature's powers were no longer any match for its own, that our gardens became benign, sunny places, though even today your grandmother's garden is likely to have some datura and morning glories and poppies. Right there, the makings of a witches' flying ointment still lurk, but the knowledge that once attended these plants has all but vanished.

Cannabis is an obvious star case of plant power. It's a plant whose evolution has led to increases in its potency, such that it produces a molecule that changes the content of our mind, but the power of a plant like the tulip or the potato, or the apple, is really no less extraordinary. And that power to affect us, to gratify our desires, is the product of a co-evolutionary process. Why do plants have these powers? Why did they evolve to produce molecules and colors and shapes that we regard as beautiful, that nourish us and poison us, that rouse us and put us to sleep, that calm our nerves or make us jittery, that help us work, that even change the contents of our minds?

I began to think about all this after a little encounter with a bumblebee. It was early in May, and I was planting potatoes, and right next to me was a small heirloom apple tree in extravagant bloom, and it was practically vibrating with the attention of the bees. One of the great satisfactions of gardening is that it doesn't occupy your entire mental field. There's plenty of room for speculation and reflection and idle thoughts and daydreams, unlike carpentry, which I've also done, in which if you daydream, you lose a finger. So in this somewhat relaxed state I asked myself a silly question, which was: how is my role in this garden similar or different than the bee's?

As I considered it, I realized we actually had quite a bit in common.

We were both unwittingly disseminating the genes of one species and not another. The bee, to the extent that he thinks, probably thinks that he's getting the best of that flower and probably also thinks he's made the decision, "I'm going for the apple." But we know that that is a failure of bee imagination. We know that what's really going on is that the apple tree has cleverly manipulated the bee into doing things for the plant that the plant cannot do for itself, namely, move its genes around the garden and around the world.

This, of course, is classic co-evolution, two species coming together to pursue their individual self-interests. So how are matters any different between the potatoes I was planting and me? I too had been seduced, in this case by a picture in a seed catalog, to plant that kind of potato. The potatoes too had evolved to gratify my desires for a certain flavor, a certain shape, a certain color, and this was their strategy to make more copies of themselves, to get their genes, in this case, from a potato farm in the Northwest to one in Connecticut.

So my premise is that some plants, at least since the birth of agriculture but actually going back before that, have had a similar co-evolutionary relationship with us. They have evolved to gratify our desires; that's their evolutionary strategy in order to get us to work for them, to move their genes around the world. The reason that plants have come to produce an astonishing array of molecules, that they've become nature's alchemists, is their immobility. What they've done is use chemicals, for the most part, instead of feet. They use molecules that either attract or repel other species for defense or as an aid in reproduction.

And those plants that managed to put to work a particular animal with a very large brain, a tool-making capability and a propensity to do a lot of wandering around the world, have done very well. If these plants were naming the phenomenon they would call it "the domestication of humans." Because of our new relationship with a few plants, we settled down and gave up our life as nomads, a life that had many advantages, was healthier in many ways and required a lot less work than agriculture. We became farmers; we cleared the world of its trees, by and large, to make these plants very happy, because a lot of these plants compete with trees for sunlight.

But to judge by our actions, we're still Cartesians when we look at the world. The genetic engineering of plants, for example, is totally imbued with the machine metaphor and the idea of the inertness of nature, the idea that we can take parts from this plant machine and put them in this other plant machine and get it to produce the same thing, in exactly the way we want. But I feel we have to understand the reciprocity in nature and begin to feel it in our bones.

A plant's eye view can help us do that. Domestication's not something you can simply choose to do. It's remarkable, actually, how few plants and animals have been domesticated successfully. They only do it when it's in their interest. They don't have to do it. A classic example is the oak tree. For thousands of years, people have been trying to domesticate the oak, but they have yet to produce that sweet acorn. I think it's because oaks already have a deal with squirrels that works for both species, and they don't need us.

I'm not someone who believes in the consciousness of plants. I don't have any evidence of it, but I do believe in their genius. They are far beyond us in sustainable solar power (i.e., photosynthesis) and organic chemistry. They have evolved differently than we have, but I don't view us as superior. And, ultimately, I'm as interested in *Homo sapiens* as I am in tulipa or cannabis.

I use plants as a mirror in which we can see ourselves, and that's because these plants are brilliant students of our desires. They have gotten ahead by figuring out what we desire. And so we have to pay attention when we notice that nearly every culture had a plant or plants that it used to change consciousness. It appears to be a universal desire. Tetrahydrocannabinol or THC (the main active ingredient of cannabis), for example, is an amazing invention of plants. It probably was not invented to get people high. The plant had its own purposes for developing this molecule, but once humans discovered the intoxicating effect it had on them, the plant's co-evolution took it down that path, until it was stronger and stronger and stronger on that path. On another path, it became a longer and stronger fiber (used industrially as hemp). Cannabis went down these two paths, both were human directed.

What we call drugs are, in many ways, the most astonishing chemicals that the plants have come up with, and the fact that plants can strongly affect our brains has changed us. I truly believe that the history of the imagination in the West has been, at crucial junctures, profoundly influenced by drug use. In the cultural realm drugs have acted as a sort of mutagen. They can scramble things, and most mutations are disasters or dead ends and do no good, but at times a mutation is wonderfully adaptive and productive.

There is a strong case to be made that many of the most important figures of classical Greece partaking in the Eleusinian Mysteries used a hallucinogen once a year, for example. So, even in the West, psychoactive plants have had far more of an important cultural role than is commonly acknowledged. That suggests a much deeper connection between nature and the human mind than our own culture is currently comfortable with. Nietzsche called Dionysian intoxication "nature overpowering mind," nature having her way with us. That's a powerful and perilous idea in the West. After all, Adam and Eve were thrown out of the garden because they sought to gain spiritual knowledge from a fruit, a tree, i.e., from nature, perhaps the drug war's first battle.

Today to take a leaf or a flower and use it to change our experience of consciousness is at odds with the values of our society, but I'm inclined to think that such a sacrament may on occasion be worthwhile just the same, if only as a check on our hubris. Plants with the power to revise our thoughts and perceptions, to provoke metaphor and wonder, challenge the cherished Judeo-Christian belief that our conscious, thinking-selves somehow stand apart from nature.

But what happens to this flattering self-portrait if we discover that transcendence itself can be modulated by molecules that flow both through our brains and through certain plants? If some of the brightest fruits of human culture are in fact rooted deeply in the black earth with the plants and the fungi, is matter then still as mute as we've come to think? Does it mean that spirit too is part of nature? There may be no older idea in the world.

The Greeks understood that intoxication was not anything to be undertaken lightly or too often. Intoxication was a carefully circumscribed ritual for them, never a way to live, because they understood that Dionysus can make angels of us, or animals, but that letting nature have her way with us now and again was a useful thing to do if only to bring our abstracted upward gaze back down to Earth for a time. And what a re-enchantment of the world that would be—to look around and see that the plants and the trees of knowledge grow in the garden still.

Wade Davis: I loved what Michael said about the idea that drugs can be catalysts of social transformation, but that most mutations are disastrous. It reminds me of an anecdote Dennis McKenna told me. When he and Terence and Jonathan Ott were at one of their hallucinogenic gatherings, Dennis poked his head up from the foliage and looked around at all these psychonauts and said, "If this shit is so good for you, why do we all look so weird?"

I also remember when I was first traveling in Colombia and other parts of South America with Timothy Plowman, who was probably the best botanist of this generation, the world's authority on coca, and an incredible poet of nature. This was in the 1970s, around the time that a book, *The Secret Life of Plants,* had just come out. It made a big deal about how you should sing to your plants and touch them and talk to them and so on, and Tim hated that book. He used to say, "Why would a plant give a shit about Mozart? They can eat light, isn't that enough?"

I never cease to wonder why we are so universally ignorant about a whole realm of life—the plant realm—that is so vital to our own well-being, and clearly to the fate of the Earth. I remember up at our fishing lodge in northern British Columbia, we had a wonderful friend, a former ambassador and a highly educated scholar, and he was lamenting the demise of educational standards in America.

He went on and on in his elegant way about how people graduated from college without knowing anything about history, and their knowledge of French poetry was non-existent and, of course, they knew absolutely nothing about the great Greek philosophers, and so on. I interrupted him and asked him, "Dick, could you please give me the formula of photosynthesis?"

He just laughed his head off, because he didn't have the foggiest idea, though that simple equation of carbon dioxide and water combining with photons of light to produce food and the very oxygen upon which we subsist is really the most important chemical formula ever to be developed in the history of evolution.

I've thought a lot about my own personal background in studying plants, because I began extremely late in life. Although I was raised in the bush in British Columbia, and had a very close affinity for the natural world, I never took a single biology course at all until my third year of university at Harvard, and that was largely because biology in an academic sense, to me, was synonymous with white frock laboratory technicians and rats that smelled of formaldehyde.

But then in my third year, I picked up on the presence at that campus of a legendary figure, Richard Evan Schultes, who, in a world of few heroes, stood out and loomed large. A kindly professor who shot blowguns in class, he kept a bucket of peyote buttons outside his door; they were available to students as an optional laboratory experiment. He was *the* man who, to a large extent, had sparked the psychedelic era with his discovery of the use of magic mushrooms in Mexico in 1938, and had made unbelievable solo research odysseys in the Amazon in the 1930s and 1940s.

He was also notorious for the dexterous manner in which he got students off marijuana convictions. By legal statute, of course, *Cannabis sativa* was then illegal, but Schultes and a small minority of botanists maintained at the time that there were, in fact, three psychoactive species of cannabis, including *indica* and one called *ruteralis*. Because of this and according to Schultes, that left the burden of proof in a criminal prosecution on the prosecutors to show beyond reasonable doubt that a ground-up bag of weed was in fact *sativa* as opposed to *ruteralis* or *indica*. And since even the botanists couldn't agree on how many species there were, it was by definition an impossible task.

Now, none of the students actually cared about these arcane points of nomenclature or taxonomy. However, they did appreciate the marvelous ability of this fifth-generation, ultra-conservative, upper-crust Bostonian (who was still loyal to the British monarchy!) to break open the courtrooms and set the children—guilty only of the sin of smoking an innocuous herb—free. At one point, he was testifying at a case in Cleveland, and the judge turned to him and said, "Isn't it true, Professor Schultes, that you simply come to these trials to increase your own personal prestige?"

Schultes turned to the judge and said, "Sir, do you think I, Richard Evan Schultes, a fifth-generation Bostonian, would come to Cleveland to increase my prestige?"

When I first went to him and said, "I saved up some money in a logging camp, and I want to go to South America like you did and collect plants," he simply looked across the room at me and said, "Well son, when do you want to go?" And two weeks later I was in the Amazon with Tim Plowman where I would stay for fifteen months. Before I studied botany in the field, a forest was just a kind of a monotone of green, interesting in an aesthetic way in its totality and its beauty, but ultimately incomprehensible and mysterious. But then, as I began to delineate phylogenetic relationships and began to understand the meaning of taxonomy and even the poetics of the Latin names themselves (which Tim Plowman recited effortlessly as if they were Zen koans), the entire dimension of that forest suddenly took form. Plants became cousins to each other. You could understand, say, the relationships between all the members of the *solanaceae* (nightshade) group and why that family was the family of choice of black sorcerers and witches throughout the world.

I began to realize the spatial and geographical wonder of the plant universe and its extraordinary evolutionary history, the marvel of the emergence of flowering plants which largely succeeded the conifers, and so on, and I just couldn't understand why plants were seen to be boring by so much of our society. I've met a lot of botanists who are boring, but I've never met a single plant that is boring. When I returned from South America, and began—against all of Tim's expectations—to actually become a diligent student of plants in an academic sense, I would be in the science library at night moaning and groaning with revelation.

The night I discovered what the Krebs Cycle* meant, I actually got thrown out of the science library at Harvard for making too much noise. The other students in my botany classes were complaining about having to memorize all these metabolic pathways, but because I was an anthropologist and had studied so many myths and kinship links, to me these were just stories and kin relationships that I committed to memory effortlessly. Metabolic pathways were like songs the plants had sung to me.

Based on my studies in both anthropology and botany, I agree with Michael that the desire to periodically change consciousness is a ubiquitous instinctive trait of the human species. It's something Dr. Andrew Weil has written eloquently about. It is something you can find everywhere, and that forces us to question why there is this kind of severe restriction on the use of these sacred plants in our society. But it's also very important to remember that the use of these plants is rooted in culture and—of the known hallucinogenic plants, and the one hallucinogenic animal we know of, in the toad family—the vast majority are from the Americas and Siberia.

There are psychoactive plants elsewhere in the world, but the use of plants to satisfy that desire has a disproportionate concentration in the Americas. In other places, peoples have found other ways to transform their spirits, through music, movement, and spiritual disciplines. And

---

*The Krebs Cycle is the metabolic sequence of enzyme-driven reactions by which carbohydrates, proteins, and fatty acids produce carbon dioxide, water, and ATP.

there are places where psychoactive plants may exist and have been used on occasion but where an ongoing consistent tradition of use is absent. Though, if you also consider milder mood transformers such as nicotine and caffeine and alcohol and so on, then it is universal.

I'm very interested in the idea Michael Pollan mentioned that agriculture was, in a sense, a strategy of grasses to get rid of trees. I am very interested in the advent of agriculture and the death of our nomadic traditions because I've spent a lot of time with nomadic peoples. I'm temperamentally different than Michael. I don't really even like domesticated plants. I don't like domesticated anything. In fact, I like to live in places where there's nothing domesticated, and in a way, it made me think those few grasses that have used us so effectively (wheat, corn, etc.) may currently be the fascists of the plant kingdom. What do you think about that?

Michael Pollan: I think gardens are very hopeful places. To me they show a mode of inter-species co-operation and reveal a history of our interests and other species' interests being reconciled in a certain way, coming together, being reciprocal. I think we need positive models like that, and that's one of the reasons that my environmentalism flows out of the garden, rather than the wilderness, which is where Americans traditionally have gone for their ideas about nature.

On the other hand, plants, just like us (and all sorts of species), expand to the limit of their ranges, and that process is stopped by the cruelest, most disastrous mechanisms. A monoculture, the ultimate success of a species, is usually a disaster. Take the lumper, the potato that conquered Ireland. It was a strain of potato that was a great success (in the narrow sense) in that it got itself planted all over Europe—to the exclusion of virtually every other potato. But then in 1845, a blight showed up that wiped the lumper out. The success of some of these domesticated species plays against biodiversity, which is nature's real insurance policy. Sometimes, like us, these domesticated species go too far and disaster ensues.

Wade Davis: I was being a little flippant in knocking domesticated plants, and I have to admit that pure wilderness is very rare. There's no place humans have gone where they didn't make some kind of impact or mold the landscape, even in the Amazon. Clearly, the hunting and gathering tradition colored most of our history, and when we talk of Neolithic shamans using medicinal plants, these were obviously plants gathered from the wild. But the rise of agriculture and domestication, once we got it down, happened very quickly. An idea that now seems so obvious to us—to encourage the growth of a seed and select for certain properties of the progeny of that seed—just took off. It's sort of interesting how it took so long, and then when it happened, it happened so quickly, and the consequences have just been extraordinary.

As an anthropologist, I'm more concerned about the *consequences* of domestication, because it clearly set into motion a lot of the processes that are afflicting us today, as well, of course, as most of the things that we benefit so much from. But I have to admit that I prefer being with nomads. I prefer to be with nomadic pre-agricultural people if I have the chance.

This discussion took place at the Bioneers Conference in 2001.

8

Women, Plants, and Culture

Kathleen Harrison

I have worked many years in the realm of people and plants and plant medicines. Twenty years ago I spent quite a while in the Peruvian Amazon working with healers who used a spectrum of plants from the most subtle to the most powerful, and then I came back into my life in California and did many other botanical projects and also raised children. It was only in 1993, when I went back to the Amazon, that I completely understood the idea of "plant spirit." In fact it was no longer an idea; it became a reality and got under my skin and changed my life substantially.

Since then, I've worked for a couple of years with Mazatec Indians in the mountains of Oaxaca in Mexico and also learned an increasing amount about the indigenous Native Californian relationship to nature and plants. I see wonderful parallels in these nature-based societies in which everything is viewed as animate and every species is a being. I think we too need to develop an intrinsic perception of this hidden reality in order to make the medicine that we grow or are given or buy even more powerful and more effective. Our culture is extremely reductionist and materialistic, so we really need to learn from models that can help us understand spirit in nature and spirit in medicine.

The word "spirit" comes from the Latin for "breath," so what we're talking about when we say that a plant or a species has a spirit is that it draws energy from the universe and expresses it in a particular form. An ancient notion in many indigenous cultures around the world is that there were primordial beings on Earth before we came along. They interacted and had relationships and love affairs and conflicts and exchanges of all sorts, and each of those beings became a species.

According to these creation stories, we humans, as complex and differentiated as we may seem, are one being, and each of the plant species that we use as medicine is also one being. I've learned from my native friends to talk to the spirits of those plant species. Whether you're ingesting a plant or growing it in your garden or passing it in the forest, you learn to talk to it as though it were a being that in its genes and in its form carries a constellation of qualities, actions and ways of interacting with us that we need to speak to, not just know about. To know about it is a step, but to speak to it and listen to it is really what makes the medicine work, because then it becomes a relationship.

That's why in many parts of the world it's women who carry the knowledge of plants and who gather the medicines. Women are good at relationship. Like all female mammals, and most particularly female primates, one of our roles is to be nurturers and doorways for life. We often hold in our hands, for our lifetimes, the job of being caregivers.

When children come through us, they are not ready to be in the world, of course, by themselves. In some species they are, but not in mammals, so we have to give them a deep level of attention and read their needs in a way that goes far beyond the verbal in order to help them survive and grow and thrive. We have to give them intense care and attention. We women have to be able to sense, on all levels, the needs of the beings around us and to use all our senses in caring for the young.

These skills translate well into gathering plants. We can look back through the entire history of hominids and see that we survived and thrived by using our senses of tasting and smelling and touching and seeing to make clear distinctions between one plant and the next, between safety and danger, between all of the different ways that the elements present themselves to us, and it was mostly women who perfected these skills.

I think it's important for women to recognize that we have become very skilled at paying attention to the natural world; these skills allow us to call nature in through us—to be food, to be medicine, to be magic, to be whatever the many forms of partnership are between the plant world and human women. Language itself can be an obstacle in this quest. It can be a sort of screen we get trapped behind, separating us from the multi-leveled reality behind it.

I love words, but our culture has bought into the idea that there's an objective reality, and that as individuals or sub-groups or sub-cultures we may have our own kind of group subjectivity, but that there's a definite objectivity that trumps that. It's important to remember that that objectivity is, in many ways, really also a cultural construct. I think we're becoming braver about showing what we know and not being held back by the inner judge that says, "Objectively speaking, that sounds crazy." I think women are often more willing to trust their subjective experiences, their dreams and intuition, and are therefore more able to develop a relationship to the plant world.

I'm encouraged by what I've been seeing. I've been speaking at herbal conferences and teaching at the California School of Herbal Studies these past few years, and many of the people who come to these gatherings are young women; some three-quarters of them are young women who are studying to be herbalists or who are called to plants in some way. These young women studying herbalism and studying ethnobotany give me heart because they're starting farther along than many of us. My generation (I'm past fifty) had to work through so much old stuff just to begin to trust our instincts, to begin to know nature, to begin to rediscover that our roots were in the earth too.

I now see many young women coming in with the assumption that they are rooted in the earth, and that they have the ability to be healers. It's a generation of healers that is starting from a very good place. I want to encourage this new generation to learn to listen to the ancestors and to the voices of all of the species around us, however humble and subtle. I'm a champion of subtlety. The subtler something is, the more you have to pay attention, and that's a good thing. It's not always the big, loud species that are the best teachers. Sometimes it's the little, quiet, humble ones.

One thing that I've realized in my travels studying people and their relationship to nature is that plants have the ability to transmit energy. Plants draw and transform earth and water and nutrients and light and make their bodies out of them. The plant beings are manifestations of these forces being woven together, and we humans have relied on them to sustain us from the beginning of our evolution.

Cultures that are close to the earth recognize the power of plants to hold and draw energy in a situation and change it in a healing way, to move it. The plant world is constantly whispering to us if we can hear it. There's been a long partnership between plants and, to a large degree, women, around the world, who know how to take plants with offerings and with prayers, and to use them to move energy. This is part of all of our ancient traditions, and we're gradually returning to these ways.

These concepts are rooted in the day-to-day way that native peoples live. When relating to the natural world, one of the most basic principles they abide by is reciprocity. When, for instance, they meet a plant and wish to take some of its body for medicine, they ask its permission and explain why they need it, and then they give it something back. On this continent what has often been given back is tobacco. It is the most sacred traditional spiritual plant of the Americas, and was apparently domesticated thousands of years ago by indigenous early Americans.

I've thought about what is most valuable to people in our contemporary culture, and I think it's time. Time is the thing that we have the least of, that we're the most jealous with, and that we can offer a plant if we want to get to know it and we want to ask something from it.

The way we can offer it time is to learn about it, sit with it, and maybe grow it. But even if you're just purchasing some of it, try to learn about that plant's world. When we use a plant, we're communicating with the entire chain of experience of that species through its evolution.

Medicine, in the traditions I've worked in, is not just about chemistry. It's about that which heals. Many cultures talk about the songs that come through the plants. If you listen well to a plant that you have solicited medicinal aid from, they say you can learn its song, and its song will be as effective a medicine as the plant material itself. That's when you've taken that plant in as your deep ally: when you can invoke its medicine without even necessarily touching or finding the plant. At that point you have access to the spirit of the medicine.

It has been part of my work to go to cultures that use sacred plants in Mexico, Ecuador, and Peru and learn their mythology and sometimes their ceremonies. These traditions and these sacred plants have to be met with total respect. I ache when I think of any of these sacred plants becoming mere commodities in our culture. The commodification of spirit is really dangerous territory, and we're not generally wise enough and open-hearted enough to take that type of medicine on our own, for casual use, without a teacher, a healer who can show us how it really is medicine. But I can certainly grow a beautiful little peyote plant, and it can be a teacher to me, even if I just watch it flower and act as its guardian. A tiny, little, delicate plant can have a very powerful spirit.

Our culture still has a strange relationship to plants. For example, it's interesting that finally, after a very long period of denouncing cannabis, we've opened the dialogue enough to talk about it again as medicine, as it has been for thousands of years to many people. Little by little, money has come forth to do studies on its medicinal use. A recent study concluded that cannabis actually relieves pain, and there are so many receptors for it in the human brain that it is uncanny.* When used with

---

*For an overview of all the studies that have been done on cannabis and psychedelic substances, the incredibly exhaustive website Erowid.org is highly recommended.

intention and gratitude and awareness, it can be a multi-faceted medicine. It is our sister and ally in many ways.

I try to keep my eye on the big picture and observe how cultures vacillate in their appreciation and rejection of powerful plants and how fear and denunciation cycle around again to an appreciation of plants. We're in a time of such fierce denunciation of tobacco right now that we find it very hard to talk about its holy and sacramental properties. A plant is in itself not evil or good. Its effect depends upon how we use it and how conscious we are, and that's true of all medicine. Unconscious use of anything is damaging whereas conscious use of anything can make it medicine. This goes for food and all the other substances that we love and hate, and yet we seem to continue to go through periods of demonizing aspects of nature that we don't understand.

Chocolate is another fascinating psychoactive plant. Chocolate pods are filled with beans that have been used as offerings (and still are) in Mexico. People give them and other plants and seeds to the Virgin Mary and/or to the spirit of a mountain or the local gods. They give them things that they know will please them (and how could chocolate not please them?). Some plants they burn as incense and some they just lay in sacred places where water wells up out of the ground. They don't necessarily bring flowers or spectacular looking things but often more subtle gifts. You bring nature to nature because it shows that you have paid attention and understand reciprocity. You don't take without giving something, and then you're always grateful for what you get. That's medicine.

This presentation took place at the Bioneers Conference in 1997.

9

Visionary Plants Across Cultures

Edison Saraiva, M.D., Kathleen Harrison,
Dennis McKenna, Ph.D., Charles Grob, M.D.,
Andrew Weil, M.D., Marcellus Bear Heart
Williams, and Florencio Siquera de Carvalho

Edison Saraiva: I am a physician and homeopathic doctor in Brazil, and
I work in the area of eco-toxicology, the toxicity of the environment as
it affects human beings, and I also do research on nutrition. I'm part of
the União do Vegetal (UDV) church. I have been drinking hoasca for
eighteen years. During this time I've lived partly in the Amazon area and
I've worked with the Ministry of the Interior of a regional area of north-
western Brazil. Throughout these eighteen years, I've been drinking the
hoasca tea, and I find it helps me to balance my governmental work
and my inner life. It is interesting how one can inhabit these two different realms of reality and maintain excellent mental health. The hoasca,
when well administered, brings the ability to transform the subtle, non-
material world into something very palpable.

Kat Harrison: I'm the president and project director of Botanical Dimensions, a nonprofit organization I co-founded for the purpose of collecting, protecting, propagating, and understanding plants of ethno-medical significance (including shamanic plants) and their lore. Part of my intent is to broaden our cultural definition of what healing is and what healing plants are, what medicine is, and to incorporate the shamanic plants into categories our culture is gradually accepting in terms of herbal medicines. Botanical Dimensions operates an ethnobotanical garden in Hawaii and is helping to establish another in Peru. I've explored these various plants and substances extensively over the last twenty-some years in several cultures throughout the Americas.

Dennis McKenna: I'm also associated with Botanical Dimensions as a research director. If you look at cultures in which shamanism has a strong tradition, almost invariably you find that that tradition is centered on the use of one or more powerfully psychoactive plants. I did my graduate research in ethnobotany in Peru in the early 1980s, studying ayahuasca, and I came away from that experience with the intuition that this was really a very interesting drug or plant complex, and that it and the methods of the traditional healers who used it were worthy of investigation from both a medical and an ethno-botanical point of view.

I encountered many *ayahuasqueros* in my field work who had used ayahuasca on a regular basis, some for most of their adult lives. Far from being what we think of as being impaired by drugs, they were actually extremely mentally well-balanced, physically healthy people who were extremely high functioning. They actually seemed to have derived a lot of benefit from the incorporation of ayahuasca into their lives.

I realized that we didn't really know anything about the pharmacology of this drug in humans. We knew a great deal about the chemistry of the plants and their effects on animals, but animal models are not really adequate to describe how these things work in humans. I had long wanted to do a biomedical study of long-term ayahuasca users, but it simply wasn't feasible to do such a study with traditional indigenous populations or jungle-dwelling *mestizos,* the main users of these substances. As a result, I had put this idea on the back burner until last year when I was invited to a conference in Brazil to give a paper on my chemical and pharmacological work on hoasca. I was invited by Edison and the União do Vegetel church—the organizers of the conference.

When there I realized that the members of the UDV could provide an ideal group to study. Many were urban professional people who could be monitored and interviewed and would give their full consent to have medical tests performed on them; they would understand the value of the research. When I broached the idea to Edison and some of his colleagues, I was a little apprehensive because they regard hoasca as a sacrament. I was worried they might feel I was being blasphemous in some way, but I found in them an extreme attitude of openness and a desire to understand their sacrament on all levels—from the biophysical to the metaphysical. They were enthusiastically receptive to the idea.

So we began working together and I thought I would come back to the United States and write a grant and submit it to the National Institute on Drug Abuse (NIDA), the government agency that funds these kinds of studies. I started to write the grant. I worked on it for a year or so, and as I was working on one of the final drafts, I was reading it through and I realized that NIDA would never go for it. I realized that the only chance to fund such a project was to find some private individuals with resources; individuals whose lives had been touched by shamanic plants and who would be willing to fund this sort of research.

Lo and behold, a couple of very generous individuals, one of whom is Jeffrey Bronfman, came forward. We picked Dr. Charles Grob from UC Irvine Medical School to act as the principle medical investigator, and we are now ready to initiate a pilot study on the biomedical effects of ayahuasca in people who have used it for many, many years.

Charles Grob: I'm a psychiatrist at UC Irvine. I have had a long-standing interest in the potential medical and psychiatric application of psychedelic drugs. In fact, that is a good part of the reason why I initially decided to go to medical school. I felt that in order to do that sort of research, I had to get unimpeachable credentials. When I made that decision to go to med school twenty years ago, I figured that in about five years the country would come to its senses and we would be allowed to do sanctioned research the way it should be done. And yet here we are, twenty years later, and we are barely now on the verge of *perhaps* being able to resume some serious research on psychedelic substances.

Most people don't realize that some thirty-five years ago psychedelic drugs were one of the hottest topics of study within psychiatry. It was widely felt at the time that they might hold an important key to really understanding how the mind works, to understanding psychopathology and to developing new treatments that could help people overcome some mental afflictions and live healthier, more fulfilling lives. I think that potential is still there. Tragically this research was blocked twenty years ago and it has not been allowed to go forth, but I think that is starting to change, though still very slowly and hesitatingly. I had submitted a protocol to do a study on MDMA (Ecstasy), which may have considerable therapeutic potential.

After a long series of drafts of a protocol, we submitted a proposal to the FDA and the decision there was very encouraging. They essentially felt that this was a worthy area of study and that basically MDMA should be treated no differently than any other drug. Perhaps we are starting to see a slight paradigm shift, so I was very excited when Dennis asked me if I wished to join him as a co-investigator on the Hoasca Project.

Ayahuasca has been used as a medicinal remedy and shamanic plant for thousands of years. It has enormous potential. Our study is essentially designed to look at both the physiology of experienced users when they imbibe the substance and at their biochemistry, particularly as it relates to neurotransmitter function and psychological effects. It is our hope that, with the preliminary data we can gather with our initial trip and work in Brazil, we might be in a better position to approach our own government, NIDA, or other government agencies, to approve and support further research here.

Andrew Weil: I am from the University of Arizona College of Medicine. I am a botanist and practicing physician. I practice natural and preventive medicine. My main interest has always been teaching people correct uses of plants and the most profitable uses of plants, and I include shamanic plants in that. I've always been interested in the healing potentials of the psychedelic plants and drugs, not just in psychiatric medicine, which is what most of the literature has been about, but in clinical medicine as well.

I have seen remarkable examples of healings from chronic pain syndromes in autoimmune disorders in connection with psychedelic experiences. I think it's a shame that physicians are denied the right to experiment with and use those drugs clinically, especially since from a medical viewpoint most of these are among the safest of all known drugs in terms of toxicity. Most of the things that we routinely dispense to patients are much, much more toxic than the true psychedelics.

I am also interested in magical plants other than psychedelics, and an example of one that I have had a long history of involvement with is coca leaf. I think the history of coca is the most flagrant example of the way in which we have gone wrong in our relationships with plants. Coca is the sacred plant of a large population of Native Americans in the Andes. The religious and sacramental significance of coca is enormous in these cultures.

If you are in the area where coca is used and watch Indians in an unobtrusive way, you will often see that, when they first begin to chew coca, they take three perfect leaves and put them together in their hand and then blow on them and whisper a prayer to the leaves before they put them in their mouths. There is a tradition of divination involving reading coca leaves in the Andean highlands, done mostly by women. You cannot obtain the power to read coca leaves in this way, they say, until you have first been struck by lightning and survived. These are just a few examples of how central coca is to the spiritual life of Andean culture.

When the Spanish Conquistadors came to the New World, their

immediate reaction to coca (and to most everything else associated with Indian life) was that it was satanic and should be suppressed. They tried to do that initially, but then they discovered that Indians worked better if they were given coca. The Conquistadors enslaved a lot of the Indians and put them to work in mines, and found that selling coca to them and profiting from the sale of it—as well as getting enhanced labor from the Indians—was in their own selfish interest. As a result, for the next couple of centuries, the only interest that Westerners had in coca was that it was a profitable tool by which they could get more work out of Indians.

It wasn't until about 1869 or 1870 when an Italian neurologist wrote an essay about coca, pointing out that it had very unusual, interesting properties, that Europeans started taking notice of it. Within a year of that, cocaine was isolated from coca, and all scientific interest shifted to cocaine in the belief that all of the active properties of the coca leaf were to be found therein in a form far easier to measure and study than in the whole plant.

We are paying dearly for that reductionistic mistake to this day. It has led to an epidemic of cocaine use in the world, created entirely, initially, by the medical profession, which handed it out as a panacea, thinking that it had no downside. Eventually the problems caused by its abuse led to a great public outcry against cocaine. At that point, the medical profession then had the same response it has had for a century or so to every psychoactive drug that it has initially mis-prescribed: it deflected any blame for its prescribing practices and labeled coca an inherently bad substance with no redeeming qualities. When you do this and a drug is banned, instantly a black market comes into existence to supply the thousands of people who have become addicted to it as a result of careless prescribing by medical doctors.

The laws against cocaine drove the safe form of that plant, the leaves, out of circulation. The leaves have medicinal properties and are not very prone to abuse, as they have low addictive potential. Thus an enormous black market in cocaine was created. Massive efforts are now made to eradicate the plant in the areas where it grows. There has been a constant, bloody, destructive war against coca growers (and the ecosystems in which they live) in the Andes, stimulated by the international narcotics control bodies with the United States at the helm. Clearly, this war suppresses Indian peoples. The chewing of coca is a very powerful component of Indian identity in these regions, so the people who want coca to go away really want Indians to go away, or they want Indians to turn into us, though the rationale is always posited in medical, psychiatric terms.

The quality of scientific research on coca's supposed harm in the Andean areas has been dismal. To cite only one example, in the late '60s, the United Nations sent a team of Canadian psychologists into Andean villages to administer standard Western intelligence-scale tests to Indians to "measure" their intelligence. They somehow concluded from the results of this culturally absurd, racist exercise in bad science that coca caused mental deterioration and brain damage, and this was then used as a further rationale to step up the war against the coca leaf.

The shame of it is that coca leaf has a whole range of interesting therapeutic qualities not attributable to cocaine alone but to a wide range of chemicals found in the whole plant. Positive aspects of the plant are lost when it is transformed into cocaine. I think of all the many cases I have looked at of magical plants and relationships that people have formed with them, I have never seen one in which it is so obvious what our culture has done wrong. The scale on which the mistake has been made and the costs of it, both to us and to the native populations that originally knew how to use this leaf, are flagrant and horrible and tragic. It would be wonderful if we could shift that.

Marcellus Bear Heart Williams: I have an adopted son named Johnny White Cloud who couldn't be here today and asked me to come to say something about the Native American Church and peyote. So far chemists have found more than fifty-nine different alkaloids in peyote. It has very complex kinesthetic, olfactory, visual, and auditory properties.

There is a long history of pre-Christian use of peyote. It came to the United States from Mexico, and Indians in the Southwest began using it centuries ago, but not in an organized church context. Different people began to investigate its use in the late nineteenth and early twentieth centuries because white settlers, who thought Indians were pretty crazy to begin with, were really worried about those crazy Indians getting even more *loco* by eating that *loco* stuff.

Early in the twentieth century, James Mooney, the historian, was taken into some of the peyote meetings. He advised the Indians that they had something really good going on. He told them: "If you want to protect your peyote ceremony, charter yourselves as a church, because before too long the BIA (the Bureau of Indian Affairs, but we say BIA stands for 'Bossing Indians Around') will try to stamp it out, and not only that, but missionaries will also come and try to stamp it out." The Indians took his advice. On October 10, 1918, at my Uncle Bob Cook's place in Oklahoma, the Native American Church was chartered.

We had used this peyote long before then, and many of the old-timers who have kept traditional ways still address the Supreme Being by the same name that they used before white men came to this continent. However, one of the main leaders who helped found that church back in 1918 had become Christianized into the Methodist faith, and while he said that we should keep our Indian motifs—the teepee and our traditional paraphernalia in our ceremonies—he had been fasting and praying and eating peyote and drinking peyote tea and he had received guiding visions about the structure of the church. It would be called the Native American Church and it would also incorporate Christian teachings. Because he was a respected leader, people went along with his vision.

In this church many miraculous things have happened over the years. A lot of psychological and spiritual power seems to evolve within a circle as we do our ceremonies, and people have been cured of serious illnesses. Many, many people have straightened out their lives and stopped drinking. It is not the peyote itself. The peyote is only used as a focal point for the power of the creator. We don't worship this herb. We acknowledge it as a gift from the Creator through which we manifest many positive forces that can help us heal emotional, physical, or spiritual illness.

Florencio Siquera de Carvalho: I am a very simple person. I have lived most of my life in the forest. I have learned a lesson that has been very important for me: I know that I don't know anything. I am a member of the União do Vegetal and I hold it and all its members very dear to my heart.

I have been drinking ayahuasca tea for more than thirty years. I leave it up to your criteria, you who are well-studied and knowledgeable about many things, to judge whether this tea has or hasn't been beneficial for me. We use this tea for mental concentration and to cure illnesses of all types. One of the illnesses that the vegetal (ayahuasca) has cured the most is the fighting amongst neighbors. It has brought people to join their hands together to be friends and brothers. For this reason, I am proud to be a member of this sacred organization, the União do Vegetal, and I don't consider this particular pride to be a human fault.

MODERATOR'S QUESTIONS AND
PANELISTS' RESPONSES

Moderator: How might we use these plants therapeutically outside of a traditional ritual context?

Andrew Weil: One of the problems that happened with early LSD research was that researchers who took LSD themselves and had positive experiences with it and understood its potential to alter consciousness in ways that could serve healing, conducted studies and published their results, but then other people who had never taken LSD and didn't understand the way it worked and thought of it as a magic bullet went about administering it in other circumstances without attention to set and setting and didn't get the same positive results. That kind of controversy in the literature, I think, scared a lot of other researchers away from attempting to use LSD therapeutically.

The point is that these drugs in themselves don't have absolute properties. They can have certain positive therapeutic properties if they are used in the right context and if the expectation of both the patient and the doctor are supportive. That is difficult at the moment to explain to many physicians and many psychiatrists who think that the magic is all in the substance. It's a matter of training people to use these things correctly. The first prerequisite is having the experience yourself and in the right sort of context so that you can transmit it to other people. There's no substitute for that.

Moderator: I was wondering if anyone who's involved with the Hoasca Project has thought of the idea of using it as a possible therapeutic tool to help rehabilitate prisoners and criminals?

Dennis McKenna: Certainly part of the objective of the project, which is a long-term process, is to look at many potential therapeutic uses. I think it would be great if you could use it to treat, for example, chemical dependency. I think it might have great possibilities in that realm, but the initial objectives of the project are more modest than that.

Basically nothing is known about the pharmacology of ayahuasca. We are trying to get some baseline data, some limited amount of information about how it acts on human physiology and on human cognitive functions. Once we have that, hopefully, out of that will emerge some suggestions, some ideas as to where to go.

In other words, the data sort of defines itself. The way that science works is that you do experiments, you get results, you look at the results, and then you try to ascertain what the next logical question to be answered is. But because of the political and regulatory climate in this country we are a long way from being able to use ayahuasca or any other traditional consciousness altering substances in that sort of human research here, especially in prisons. That doesn't mean that that sort of thing can't be pursued in Brazil where religious ayahuasca use is now legally recognized or in other countries where there is more openness to such research.

Marcellus Bear Heart Williams: We in the Native American Church really have to hurdle the Food and Drug Administration to get peyote in to our church members who are in prisons, but our ceremonies have helped reform quite a few criminals. I'll give you an example. Roland Hague was a notorious troublemaker who did drugs, stole cars, robbed banks and went to prison. Even his mother had given up on him. But when he came out, somebody brought him to the church, and he never went back to drugs or crime. He is now a very respected member of the Cheyenne tribe. There is no doubt that it can turn things around.

Edison Saraiva: In Brazil we are already working all the time with ex-prisoners among our members in the União, getting excellent results with their rehabilitation and social re-integration, and we would like to be able in the near future to also work with prisoners while they are still behind bars.

Moderator: I'm wondering what you think will happen with the use of ayahuasca analogues in North America. I've heard people have discovered high amounts of DMT (dimethyltryptamine—the most active hallucinogen in the ayahuasca brew) and harmaline (the other main chemical component of the tea) in a variety of plants that are indigenous to our region.

Kat Harrison: It's true that upon hearing about ayahuasca in the last decade or so a number of inquisitive North Americans have looked into what might be here in our territory that would have the same components. There is a lot of searching and experimentation going on with several plant species. I think some of the amounts of active components that have been reported are exaggerated, but people could probably come up with a contemporary North American botanical analogue to ayahuasca, and what would happen with it would really depend, as always, upon the attitude people prepared it with and how they approached the experience.

Such people would be wise to model their use on the indigenous traditions that have a long history of using these substances to build respectful, reciprocal relationships and to achieve healing on many levels. That's the reason that I have spent so many years studying these cultures and traveling to experience the way that different indigenous people use their shamanic plants. If we use these sacred plants, we have to do it right.

This discussion took place at the Bioneers Conference in 1992.

AN UPDATE FROM DR. GROB

Since the discussion in the previous chapter took place some time ago, the editor requested an update from Dr. Charles Grob, who is perhaps the most important current researcher studying the effects of the substances that are the topic of this book. The update describes the research he has done in the intervening years and his thoughts on the current state of this area of inquiry. Dr. Grob was gracious enough to send in the note below:

> Since my initial remarks were made, almost fifteen years ago, much has changed and much has not. In the years following 1992, our research team, including Dennis McKenna, Jace Callaway, and Glacus da Souza Brito, conducted a comprehensive Phase 1 study of ayahuasca in the Brazilian Amazon. In May and June of 1993 we worked with the UDV syncretic ayahuasca church in Manaus, the capital city of the state of Amazonas. Fifteen randomly selected volunteers, with at least ten years of experience taking ayahuasca as a psychoactive sacrament within the structure of the UDV, were medically and psychologically evaluated using state of the art mea-

sures. Each of these subjects was administered an experimental ayahuasca session, with serial measurements of heart rate, blood pressure, and electrocardiograms, as well as blood samples taken every
twenty minutes for four hours from an indwelling intravenous
catheter for assays of neuroendocrine secretion (prolactin, cortisol,
and growth hormone) and for pharmacokinetics (of harmala alkaloids and dimethyltryptamine). Baseline blood samples were also
taken—from the ayahuasca subjects and fifteen matched controls
who had never taken ayahuasca—for analysis of platelet serotonin
receptor density. Baseline psychiatric assessments and psychological
testing, including neuropsych measures and personality tests, were
also conducted in both groups.

Findings from the Hoasca Project have been published in
the medical and scientific literature.* Overall, the long-term
ayahuasca-using subjects were assessed as being in very good
medical and psychological condition. Indeed, their performance
on neuropsychological and personality measures were healthier
on average than the non-ayahuasca experienced controls group.
One particularly surprising finding was the up-regulated platelet serotonin receptor systems in UDV subjects, which might be
the neurobiological basis of ayahuasca's observed antidepressant
and mood regulation effects. Many of the subjects had personal
histories of remarkable transformations following initiation into
the UDV ayahuasca religion, including achieving lasting sobriety
from chronic alcoholism and drug abuse, along with significant
reduction of associated psychopathologies.

Another study of ayahuasca use in Brazil that I conducted in
2001 with colleagues Marlene Dobkin de Rios and Dartiu Xavier

---

*Psychopharmacology 116 (1994):385–387; Journal of Analytical Toxicology 20
(1996):492–497; Journal of Nervous and Mental Disease 184 (1996):86–94; Heffter
Review of Psychedelic Research 1 (1998):65–77; Journal of Ethnopharmacology 65
(1999):243–256.

da Silviera evaluated the psychological health and cognitive abilities of forty adolescents—who had begun to participate in UDV religious ceremonies with their parents, after they had entered puberty—and forty matched controls who had never taken ayahuasca. The ayahuasca-exposed adolescent subjects scored comparatively well and did not appear to have been adversely affected by their experiences. On some measures, the ayahuasca group was healthier than the controls, particularly in regard to being at a lower risk for alcohol and substance use. It is, in fact, the UDV's position that regular participation in religious ceremonies prophylax their teenage children from engaging in risky drug and alcohol experimentation.*

In the early 1990s I also conducted the first FDA-approved study evaluating the effects of MDMA in normal volunteers. Efforts to extend our Phase 1 studies to examination of MDMA as a novel treatment of psychiatric conditions, however, were not successful. Owing to both the rapid escalation and sensationalized use of "Ecstasy" by young people and the associated concern that MDMA could cause brain damage, by the end of the decade it was no longer possible to get an objective hearing of the drug's potential for treating even otherwise intractable disturbances. Although I have not pursued further efforts to gain regulatory approval for MDMA treatment research, I have published an extensive review, *"Deconstructing Ecstasy: The Politics of MDMA Research,"* that critiques the seriously flawed investigations which led to hyped and exaggerated claims of neurotoxicity that have impacted drug policy.[†]

In the early 2000s I applied to the FDA for permission to use psilocybin, an alternative psychedelic with a much less provocative public profile than MDMA, in the treatment of anxiety associated with advanced-stage cancer. Groundbreaking research with

---

Journal of Psychoactve Drugs 37 (2005):123–144.
† *Addiction Research* 8 (2000):549–588.

this patient population was conducted several decades ago by such notables as Stanislav Grof, Walter Pahnke, and Eric Kast using the prototype hallucinogen, LSD. Although their preliminary results were extremely encouraging, demonstrating marked reductions of anxiety, improved mood, enhanced quality of life, less pain and diminished need for narcotic pain medication, the final active clinical research program was shut down by the early 1970s.

Thirty years later, we were successful in gaining full regulatory approval for psilocybin (somewhat milder, shorter in duration, and more easily controlled than LSD) treatment in these patients. Our investigation, at Harbor-UCLA Medical Center, is on-going and thus far appears to corroborate many of the findings reported by earlier researchers.*

An improvement over the previous two decades when no human research with psychedelics was permitted in the United States and virtually none in Europe, the past fifteen years have witnessed the resurrection of a nascent movement of psychedelic research. In the United States, besides our psilocybin treatment project at Harbor–UCLA Medical Center, a pilot investigation of the use of psilocybin to treat severe refractory obsessive-compulsive disorder at the University of Arizona School of Medicine and a study administering MDMA to patients with chronic post-traumatic stress disorder in Charleston, South Carolina, have been conducted. In Europe, the two most active laboratories using advanced neurobiologic and brain-imaging technologies in normal volunteer subjects are examining the effects of psilocybin and MDMA (in Zurich, Switzerland) and ayahuasca (in Barcelona, Spain).

In spite of new precedents demonstrating the feasibility of conducting approved human research with psychedelics in recent years, the scope and degree of participation by the medical and psychiatric

---

*Subject inclusion/exclusion criteria and project contact information is available at www.canceranxietystudy.org.

research communities have remained limited. But progress is being made, as the further we evolve from the tumultuous cultural wars of the 1960s, the greater the likelihood these fascinating and potentially valuable substances will be given the fair and objective hearing they deserve.

The potential of psychedelic research remains vast and largely unexplored. This research may improve our understanding of central nervous system function and malfunction and may lead to new psychiatric treatment models, particularly for conditions that do not respond well to conventional therapies (i.e., drug and alcohol abuse). As new investigators develop their research programs, it will be important that they be aware of the clinical and political lessons learned by previous generations of psychiatric researchers.

Furthermore, to optimize the safety and effectiveness of their treatment, it will also be critical to appreciate and incorporate many of the lessons learned from the shamanic models of plant hallucinogen use as practiced by indigenous cultures over millennia.*

---

*Journal of Psychoactive Drugs 21 (1989):123–128; and Heffter Review of Psychedelic Research 1 (1998):8–19.

10

An Ethnobotanist and a Mycologist Discuss the Rewards and Risks of Sacramental Plant Use in a Modern Context

Kathleen Harrison and Paul Stamets

Kat Harrison: In the 1960s, when the use of consciousness-altering drugs really made an impact here, it was a real revolution, with all the creative chaos that comes with revolutions. I came of age then and I was immersed in that revolution, and to this day I say thank you to LSD because of the new windows it opened in our very narrow materialistic culture. It was an entry point, and it has its own clarity, its own character, which I still acknowledge.

But over the next thirty-five years I learned a lot by observing indigenous people who honor and use sacred plants in a very appropriate way. It has been said about the conquest and colonization of the Americas that the Europeans eagerly grabbed the land but forgot to ask for the operating instructions. A significant part of those operating instructions had traditionally come through the sacred plants that had been here and been in use for eons before the Europeans arrived. We still haven't really asked for the operating instructions. We didn't know there were any, and this has led us down a trail of large-scale destruction of our ecosystems.

I just came back from the Ecuadorian Amazon. I'd been down there a month, and I can't tell you how disturbing it is to see hundreds of roads and big, black pipelines ravaging the rainforest, and more are being built every day in all directions. A road into a rain forest marks the death of that area and its traditional people's way of life. The people there are still trying to figure what's going on. They're still taking ayahuasca and looking into their dreams and visions and looking for guidance, but their world is being ripped apart.

We get disturbed about development here in California, but as thoughtless as a lot of it is, I've been stunned by the places on the planet I've seen recently that are changing and losing their vitality and their diversity so much more quickly than we are here. It's getting harder and harder for me to transition back into my level of relative comfort and abundance and my everyday assumptions when I come back.

The ancient sacred plants can offer us a deeper kind of wisdom about the places that we live in and how to live in them. They can show us that all things are alive and constantly talking with each other, and that we need to honor those conversations, even if we can't understand them. Part of the operating instructions for how to do that can be transmitted by these plants and fungi, so we need to listen to them and to the people who have been using them for so long and who have developed the tools to access and interpret their wisdom.

We need to help those native voices be heard, especially because their lives and cultures are under siege. For a number of indigenous cultures their main sacred plants and the rituals associated with them form the *axis mundi,* the core basis of their worldviews. These are peoples that have survived a long time through incredible challenges, and they relied heavily on plant-induced visions to help them assimilate and organize the enormous amounts of information in the ecosystems all around them. These medicines help them see the deeper connections between things, achieve an overview, and receive flashes of insight. These medicines also, of course, help heal people.

I have done extensive fieldwork with the people who know psilocybin mushrooms and *Salvia divinorum* and morning glory seeds most deeply, the Mazatecs of Oaxaca, who are recognized by other Indian groups in Mexico as being the masters of plant knowledge. They have impressed upon me that their prayers go much farther and are much clearer and stronger when they ingest one of these allies—strictly following the proper rituals—and ask to speak to it, and ask it to carry their prayers higher.

They differentiate some ten to twenty related species of local psilocybin mushrooms, and they assert that each has a different character best suited for a certain type of job, a particular quality of message or a type of healing. I'd say half their prayers are prayers of supplication: "Please help us; we're so hungry" or "Please help my sick daughter" or "Please change my luck for the better." They do lead incredibly poor, difficult lives, barely eking out a meager existence growing crops on steep hillsides, but at least half their prayers are prayers of gratitude for being alive. I have found that this is universally true among peoples who use these types of plants throughout the Americas, from the indigenous peyote-using cultures in North America to the ayahuasca-using tribes in the Amazon.

They use these allies to go beyond a daily perspective and achieve a larger view, to step back and ask: "What is my work here? Am I on the right path? How am I doing with my relations, my community, and the larger world?" We so-called Westerners from more privileged societies now have access to a great gift. We can be eco-tourists and go experience botanical entheogens such as ayahuasca (sometimes it even comes to some of our greenhouses and living rooms), but part of the charge of receiving such an incredible gift is that, collectively, we have to take it very, very seriously. It can't be just about our personal growth and individual concerns. We too must look at the bigger picture, and must each take on the charge of what we are going to contribute.

Paul Stamets: I grew up in Ohio. I was fourteen years old when I discovered a book that my brother John, who had gone to Yale, had left behind. It was *Altered States of Consciousness* by Charles Tart. I was totally fascinated. It was a series of essays on the use of psychoactive plants. I lived in this very religious, dry town, and I did not fit in. I came from a family of scientists. It was normal, I thought, to have a laboratory in your basement, because we had one. So, for a while, when I'd go over to someone's house, I'd ask "Where's the laboratory?" Invariably, they'd show me the bathroom. "No," I'd say, "Not the lavatory, the laboratory . . ."

My brother had left his little kid brother a fully equipped laboratory. I made some great explosives. Fortunately I did not blow up the house. I grew more cautious after I had one very bad scare and did blow up one room. I lent the Tart book to a friend, and his parents discovered it and burned it. I was amazed anyone would burn a book. At one point some drug enforcement people came to a local church for an educational seminar to alert the public about the dangers of drugs. That was the first time in my life that I saw all these plant entheogens. They were on display, laid out on the table. The law enforcement officer finally realized after about fifteen or twenty minutes that this kid was much too interested in the subject and escorted me outside.

The first humans came to North, Central, and South America some fifteen to twenty-five thousand years ago (the exact time is a source of much debate). It is the most recently populated continent, so the use of plants and mushrooms indigenous to the Americas has a relatively short history compared to elsewhere in the world. However, the Native Americans' knowledge in this domain was light years ahead of what the Europeans knew. Christianity really suppressed this type of knowledge of the natural world, so the Europeans who came here were profoundly botanically and ecologically unaware compared to the Indians. When the first Europeans arrived, the Native Americans saved the Europeans from starvation and showed them how to use some local foods and medicines. Little did they know the devastation these Europeans would ultimately visit upon them.

Some of the indigenous peoples here really knew how to listen to the intelligence of the natural world. I think plants and mushrooms have intelligence and they want us to take care of the environment, and they want to communicate that to us in a way we can understand. When I use these mushrooms and other compounds, I get the message that the planet is in trouble, that we are approaching a huge catastrophe and that we're all in this ship together. I get the sense that all these spirits are speaking to me, that the planet is calling out to us, asking us for help, to control our consumption, waste, and pollution. But we are so incredibly busy in our culture, we don't know how to sit, be silent, and listen.

For some of us though, these mushroom and plant allies help us pause. They stop us in our tracks. We set aside sacred time to reflect on what really is important in our lives and in the larger world. They also open us up to other aspects of ourselves. I'm definitely an alpha male type, but mushrooms have taught me how to rediscover some of the feminine aspects of wisdom. I think a lot of men in our culture would really benefit from that. When I use these mushrooms, I feel like water. I feel fluid. I realize I have an aqueous body that's tied to the ocean and to the entire planet.

Even though we of European ancestry are recent arrivals here, if you were born on this land, you're part of this land; you are in some sense an indigenous person. We aren't the original indigenous people, but nevertheless we have a responsibility to be good custodians of our environment, to try and remedy the fact that the greatest natural disaster ever to hit this planet has been *Homo sapiens*. I do lots of lab work. I grow mycelium in media flasks and liquid cultures. I have seen over and over with mycelia in cultures that when an organism exceeds the carrying capacity of its environment, it crashes to extinction. And that is close to where I think our species is at right now.

I know a lot of software engineers and successful Internet entrepreneurs, and the ones I've met use psilocybin mushrooms. They work with very complex systems, and they say the mushrooms give them sudden insights that tie patterns together and help them achieve breakthroughs in their work. It's strange when many in the technological intelligentsia of society use plant and mushroom compounds to help advance computer science, biology, and other fields, and yet our government criminalizes their use. We have to have a paradigm shift, so that we *can* recognize the fact that we're all in this boat together and that we have tremendous possibilities of gaining knowledge from these plant and mushroom entities as long as we learn how to listen.

There's a mushroom indigenous to the northwest called *Psilocybe baeocystis*. It's become exceedingly rare. Two years ago, my partner Dusty and I were at our favorite rhododendron garden and I said, out of the blue: "Dusty, I haven't seen *Psilocybe baeocystis* in ten years. I really want to find this species again." Two hours later I was photographing some other mushrooms, I noticed some *Psilocybe baeocystis* an inch away from my camera. We picked the mushroom (all my work is covered by a license from the Drug Enforcement Agency, so this was legal) and took the specimen back to our lab. It's one of the few species that turns white, not brown, when it dries. It's an unusual species and I was really excited to watch its color change and to photograph it.

And then just two days ago, Dusty found *Psilocybe baeocystis* growing beside the gutter right near one of our buildings. How did this happen? Mushroom spores hitchhike on us and use us to propagate themselves and we had unwittingly carried spores to our house on our shoes or clothes. I bet most of us are carrying spores and pollens of all types around nearly all the time. This is a directed evolutionary response; an intelligent response of plants and mushrooms seeking to use or even be allies with humans.

I think native peoples understood this a long time ago and, of course, it's not just humans who are affected. I've studied bears a lot, and bears are very involved in spreading polypores to trees in forests. But we humans are the most populous bipedal organisms walking around, so some plants and fungi are especially interested in enlisting our support. I think they have a consciousness and are constantly trying to direct our evolution by speaking out to us biochemically and ecologically. I think we just need to be better listeners.

Kat Harrison: To pick up on Paul's comments: life seeks to proliferate, to reproduce itself in whatever way it can. Plants and fungi in their very different ways find creative survival and propagation strategies, including piggybacking on humans. I've been very interested in observing how the sacred plants we're discussing have propagated themselves *culturally*, expanding into new societies in this last quarter century. Ayahuasca was unknown in our culture to all but hardened ethnobotanists twenty-five years ago. I was blessed to find myself in the Amazon in 1976 and to have spent many months with Indian healers in the Peruvian Amazon, learning about a lot of plants and the "ayahuasca complex" (the brewing together of two or more unrelated plants that are not very active individually but powerfully psychoactive when combined). Hardly anyone back here had heard of it. Now it seems to be on everyone's lips, so to speak, at least in certain circles.

This ayahuasca tradition has been around for perhaps millennia, refining itself and taking a slightly different form in different tribes and ethnic groups along rivers throughout the Amazon areas of Colombia, Venezuela, Ecuador, Peru, Bolivia, and Brazil. It has been the largest plant-based religion in the world, but it had been limited to the Amazon basin. Now it has begun to spread out into more of South and North America and into Europe. That activity is, of course, entirely underground, but most of the richness of life happens underground.

I think it's an interesting question to ask: "Why is this combination of plants suddenly sending its tendrils up to us?" The bigger of the two plants involved in the ayahuasca brew, *Banisteriopsis caapi*, is a serpent-like vine that twists and turns in a quadruple helix as it grows up into the canopy of the forest. Its ally is a little shrub related to coffee, and these two together make this incredibly magical, purgative, healing, visionary tea that is now impacting our culture and our way of seeing and being in the world.

The other natural entheogen that's had a major cultural impact on us is *Psilocybe cubensis,* one of the more than a dozen species that, as I said, are used by the Mazatecs. They call it San Isidro, named after Saint Isidor, the patron saint of labor. They take this species when they want to get a good piece of work done, say when they want some inspiration from the gods on how to plow a new field on a vertical hillside or on how to start a little roadside business. Almost no one knew anything about these mushrooms until Gordon Wasson went down to Southern Mexico and "discovered" Maria Sabina and the use of psilocybin species as sacraments among the Mazatecs. He wrote about his expedition in *Life* magazine in the late 1950s.

This public revelation was very disruptive to the Mazatecs' tradition, but it fed our hunger for knowledge of a different order. So here too these fungi have spread into our culture and had a major impact, starting in the late 1960s. It's fascinating to look at the propagation strategies of these life forms from their point of view and not just from ours. A kind of cross-pollination seems to be occurring with our fast-paced global culture and these plants. Human beings love to move plants around. We've been doing it the whole time we've been exploring the world and colonizing it, but we're doing it very, very rapidly now and also propagating the cultural information that attends these plants.

I do, however, feel the need at this point to bring up the shadow side of experimenting with these plants. I don't think they should ever be taken lightly, and I don't think they're for everyone. Anyone who takes them will have a very hard time now and then, and will most likely have some scary inner encounters. These are tricky teachers. I encourage those who are certain they want to experience these visionary states to turn to indigenous practices and guidance. The indigenous traditions that have used these plants for generations know how to explore these realms. I also encourage people to read. There's an extensive body of deeply informative literature that can be of great help in preparing for such experiences. Approaching these substances recklessly guarantees that there will be some casualties.

Indigenous specialists who are accustomed to dealing with the interpenetration of different levels of reality know that you have to have a really good bag of tricks to navigate in these altered states, to face the shadow and entities that take various forms, as animals or different kinds of beings that challenge you and try to trick you and pull you in. Even if these might be manifestations of your own mind or of the collective psyche, they nevertheless have to be dealt with as real forces, real beings, real spirits. That's a crucially important part of the work.

One becomes profoundly vulnerable in these states. In fact, it's necessary to be vulnerable in order to be creative in that spiritual realm, and that's why psychic protection plays such a big part in the shamanic tradition. Not infrequently the Amazon shamanic practice devolves into psychic warfare with other shamans or with other invisible forces, and becomes darkly negative.

There is no doubt that in the indigenous cultures that have dealt with these medicines for a long time, it's not always love and light. Those in our culture who choose to explore these realms need to take great care of the "set and setting." It is crucial to be in a place where you feel totally safe so that your consciousness can relax and go inside. We can emulate all those traditions that create a circle of psychic, spiritual protection by a variety of techniques before doing any important inner work. Altars or really important objects that carry some power and are grounding can be very helpful.

It's really important to have ways to remember that you are a being on this planet and that your feet are rooted in the ground. Your consciousness may get to explore the clouds but you don't want to stay up there permanently. Australian aboriginals use a psychoactive plant called *pituri* that contains nicotine. Before going into a trance they sit near a tree and they visualize a thread extending from their mind to the tree. The thread anchors them to that tree. They know from experience that this will permit them to leave their body and to return safely. That is just one example of a traditional magical technique that permits consciousness to soar into higher states but acknowledges how risky that can be and therefore creates safeguards.

Having a trusted guardian present or nearby if one is on a solo journey is also a good idea, and many traditions use the invocation of ancestral protective spirits to watch over those who venture into these dimensions. There are also physical techniques that have been developed over the ages. Breathing is the key to keeping whole, as yogic and other traditions teach, and drinking enough water is essential. Some modern inner explorers find that ending their journeys with a pipe of excellent cannabis is very grounding and clarifying. This is certainly not traditional. Cannabis is of Asian origin, but these New World and Old World plants have met in the modern world, through us, and perhaps they are allies now.

In any event, those who are serious and prepare themselves correctly are much more likely to return safely, still sitting in the place they started, after having learned a lot. Ultimately, wholeness is the best protection—being whole, making sure our children are whole, allowing our elders and our peers as they die to be as whole as possible. If we create situations that are a nurturing ground for this story to keep on unfurling, while we go from spirit to matter to spirit again, there is no need for fear.

Paul Stamets: Yes, it really is important to treat these powerful tools with a lot of thought and respect. It's important not to overdo it. When I was younger, I was voracious for experiences of altered consciousness, but after many experiences I don't need to go back to it nearly as frequently as when I was forming my personality and trying to know my soul.

The entire context of the experience, all of its aesthetic and spiritual dimensions, needs to be aligned. I think it's important that you have a story. Buying mushrooms in a little plastic bag is not much of a story. Going out into the woods, learning what *Psilocybe semilanceata* looks like in the fields of Oregon and Washington, and having an experience finding your mushroom ally, that's a more satisfying story, a much richer experience. The path itself is as important to the overall experience as the final destination.

Mushrooms are not that hard to find. In the Northwest and western Oregon, and northern California, if you just get a load of woodchips from the local utility company, mushroom spores are often already in the woodchips. In some state parks I've seen *Psilocybe azurescens,* the most potent psilocybin-active species in the world, growing plentifully next to where all the Winnebagos park, with those folks who find a numbered parking slot, put up a satellite dish, turn on the television, and call it camping. These people have no idea what's growing all around them, but, as surreal as it sounds, there are places where Winnebagos are an indicator species for *Psilocybe.* Granted, that's not quite as satisfying as penetrating deep into the wilderness to find your ally, something I find truly rewarding, but it is a story, of sorts.

Above all though, never neglect safety. If you are absolutely committed to this type of experience, let somebody you trust totally know what you are doing, where you're going to be, and what options there are should there be a problem. I love being in the wilderness and have a lot of experience in it, but if somebody's not skilled in hiking and has never gone into the wilderness, obviously that is not something you should ever try on your own.

Kat Harrison: I think what Paul said about the source of a sacred medicine is very important. It certainly is among native people. Before using a plant, they want to know exactly who has grown it or collected it in the wild, and they want to be sure it's someone they know very well and whose intentions are pure. But it goes further than that. Let's say I'm a Mazatec shaman and my best friend was in the highlands and found the species he was looking for and collected some and brought them back, but along the way, a strange incident occurred. In that shaman's worldview, the fact that something strange had happened was considered a bad omen and the mushrooms would be deemed tainted; no one would take them. Indigenous masters of this type of botanical wisdom consider this kind of medicine so vibratory in nature and so absorbent of human intention and events that they take the history of a plant's origin very, very seriously. I recommend we follow their lead.

In the case of ayahuasca, these are often not wild forest plants, but are grown in the shaman's garden. Who grew these plants? How pure were the intentions of that shaman as he grew, harvested, and brewed the plants? Every single brewing is a different recipe on a different day in a different pot, under a different sky. All these factors combine in an incredibly intricate mix to affect our experience. This is not something you should just get out of a bottle from somebody and not know where it came from or what happened to it along the way. It's not just a drug. It's something else.

This discussion took place at the Bioneers Conference in 2000.

11

Plant Spirit

Kathleen Harrison, Jane Straight,
Dale Pendell, and Paul Stamets

Kat Harrison: The plant spirits, as I've come to see them, have broadened their territory in my attention throughout my life, but I got a good start early on. I had a naturalist father and a gardener mother who really taught me to look at plants. I am also a botanical illustrator, and I really believe in direct, detailed observation as a way to learn about the essence of any plant or any aspect of nature. I encourage a very intimate look at the physical bodies of plants and fungi as a way to learn from them. Initially it has nothing to do with ingesting them or owning them in any way. It's pure observation. All the ancient indigenous wisdom about the environment begins with paying very close attention.

Those interested in consciousness-altering plants tend to focus on a handful of species, but the entire botanical and fungal realms are teachers. There are different degrees of teachers, and some of the best teachers are the subtlest ones that don't hit you over the head but require that you pay closer attention. All cultures have, of course, had to pay this degree of attention and homage to the species that they found themselves living with, so we all have in our genes and in our bones a history of this kind of intimate knowledge of nature. It's only in fairly recent times that many of our ancestors lost that connection, but it's really still there for everyone to find. It's just under the surface of things.

There have been and still are many approaches to recognizing this "plant spirit" in different indigenous cultures, but they are all characterized by an attitude of deep reverence. They all view each species as possessing a distinct spirit or of being a spirit that has dressed itself in matter and taken on a certain form and appearance and chemical signature. All of these approaches recognize the importance of talking to and listening to these plants and asking their permission to be used as food or medicine before harvesting them. That's universally true.

Furthermore, all plant and animal spirits, all species, are understood to be in relationship to each other, and we humans are just one of those many, many species. And those relationships can get very complicated: fierce, loving, jealous, cooperative, or antagonistic, for instance, not unlike relationships among people.

Jane Straight: Over the years I have become the keeper of many sacred plants, and I hold the position with honor. All plants have spiritual qualities, but, in my experience, certain specific medicinal species contain some of the most profound energy. The land where my family dwells is considered by many to be a healing sanctuary, drawing plant teachers and healers from many parts of the world. Though my gardens are sometimes wild and unruly, many of these powerful visitors somehow end up at my place. I believe they are summoned by the spirits of the other plants already in residence.

Most indigenous peoples take for granted the connections between the physical, psychological, and spiritual aspects of health and healing. We in our culture have recently been rediscovering these links. A few of us have been fortunate enough to be able to travel the world to immerse ourselves in a variety of different cultural environments, including indigenous societies, and to experience firsthand the parallels one can find between many of the planet's healing modalities. Just about all of them make extensive use of plants and express deep spiritual links to the Earth.

Plant spirit is easy to tap into once it is acknowledged and the navigation to it learned. I find it helpful in my own relationship to plants to articulate an intention first, then lift the veil to their world, speak from the heart, and listen carefully. Sometimes a little guidance is necessary. Most other cultures have always had rituals to invoke plant spirit. Sometimes these are elaborate, but often they are quite primal, even crude.

When students from the local herb school come to my garden and nursery, more often than not at least one from each group will have a powerful experience and discover an especially profound connection to plant spirit. I try to help bring this to the fore by leading them in a visualization that can encourage the interaction between plants and humans or by leading them in making subtle vibrational remedies with plant materials or with other methods. I have found that these types of exercises have been great for exploring metaphysics in the garden.

I'll share a personal plant spirit story. I recently threw my back out. I was in agony and unable to get up. As I began crawling I heard a motorcycle coming down the path towards me. It was a dear friend who quickly sized up my condition and helped situate me in a warm, safe location. He found my dried opium poppy heads and proceeded to make me a strong tea. I had no trouble drinking the elixir; it turned out to be quite palatable, and I trusted my friend.

The *Papaver somniferum* spirit lovingly caressed me into her cloud of sweet dreams, creating a safe haven for me to leave my body. That's when the real healing began. I could no longer feel any physical sensation and relaxed completely, allowing gravity to realign my spine. I lay there for what seemed a blissful eternity while the opium spirit dusted every cell with clear light. It was an extraordinary experience that left me feeling extremely refreshed. Living harmoniously with plants is truly a divine dance.

But I understand that plants are complex beings and can definitely be misused. This issue often comes up for me in the business of selling plants. If it does not feel appropriate to let a particular plant go off with someone, I'll find a way to manipulate the situation so I can avoid selling it to that person. I have even, on a few occasions, tracked down a plant that someone had purchased because, upon reflection, I didn't feel right about it. A prime example of a tricky plant is *Nicotiana attenuata,* commonly referred to as "Coyote Tobacco." It carries the spirit of the trickster coyote in North American Native mythology. I know exactly where these plants are in my garden, but on certain days, for whatever reason, they literally become invisible to me. It is a strange phenomenon, to say the least.

"Showing and telling" is a great way of engaging youth in the study of drug plants and plant spirit. Students who smoke tobacco seem genuinely interested in seeing what an actual tobacco plant looks like, and most students really enjoy ethno-botanical stories related to psychoactive plants. In general our culture does not honor the spirit in plants, or the sacred in almost anything.

Examining the cultural and spiritual contexts from which drug plants originate can really provide a lot of insight for young people and help steer them away from irresponsible behavior. Honest information and honest communication about these plants is what's most helpful to young people, but some young adults, even those with knowledge of plant spirit, lose sight of the path. Some have to experience a "poison path" before they find their way home.

Dale Pendell: Some years ago, I took on the project of investigating psychoactive plants and substances from a poetic standpoint, that is, initially, by their specific effects on my own being, psyche, and physical body, and then on my poetry. My hope was that maybe something in how these different plants would affect my writing style would reveal some signature of the ally, the spirit within each plant.

The project grew from there and I also began looking at psycho-active plants in different plant families—historically and culturally—tracing the roots of their use in the Western tradition. A German toxicologist named Louis Lewin divided psychoactive plants into five groups, which I sort of follow. I've adapted it somewhat. I've put tobacco in its own group because it's both a poison and a medicine. It's a sacred cleansing plant. Some shamans call it "the muscle." It's a great healing plant, and yet we also know it's an incredible killer in our own society and highly addictive.

I became fascinated with this ambiguity one finds with many important sacred plants: they're both potentially great healers and dangerous poisons. I ultimately called it "the poison path." Nearly all the major consciousness-altering plant families in Lewin's classification have this dual nature: the *excitantia* (stimulants), the *euphorica* (including the opium poppy), the inebriants such as alcohol, and the *phantastica* (the hallucinogens). All these substances and plants have been considered sacred at some time and place in history and had rituals built around them.

These sacred plants seem to facilitate moving between boundaries, the boundaries between worlds; between the world of humans and the world of the gods. Studying the cultural history of these plants provides deeper insights into our history. For example, the most widespread psychoactive plant of our contemporary culture and the one we are the most addicted to is coffee, and in the history of coffee we find the whole story of mercantilism and the birth of colonialism. Coffee was intimately tied up with the slave trade (as were tea and chocolate) and it fueled the Enlightenment. Voltaire drank fifty-five cups of coffee a day.

Paul Stamets: I've been using mushrooms for a very long time, but I still find myself constantly humbled by them. They tend to show me things I don't expect. I've learned that the moment you begin to boast about your skills, they are quick to remind you that you are quite inexperienced. I got into mushrooms initially because I was interested in those with psycho-activity.

Later I got very involved with edible and medicinal species. But something struck me long ago in studying the taxonomy of the genus *Psilocybe*. I am a specialist in taxonomy, especially of this group. I was quite amazed to discover that the vast, vast majority of species that are psilocybin-active are connected with humans and human activities that are associated with taming the land and thereby creating trails of debris: chopping down trees, breaking ground to create roads and trails, and domesticating livestock, for instance.

Humans have the dubious distinction of creating the largest debris trails of any organism on this planet. When humans migrated here some twelve (or fifteen or twenty?) thousand years ago there were no people in the Americas, but after the migration of people from Eurasia over the Bering Straight, and as these humans created debris trails, many of these psilocybin-active species came out of the landscape and proliferated. Their evolution and the evolution of the human species in these migratory patterns were intimately tied.

When I have ingested psychoactive plants and mushrooms, there is one message emanating from this world of plant spirit consciousness that comes to me loudly and clearly virtually every time. That message is that we are part of an "ecology of consciousness," that the Earth is in peril, that time is short, and that we're part of a huge, universal biosystem. And I am far from alone. Many people who have taken these substances report receiving the same message.

These plants are messengers from nature, in a sense calling out to us and seeking to enlist our support. Nature speaks to us through these plants. When we take them, we usually take time away from our usual immersion in the noise of our daily lives. They can reach us if we can get away for a while from the mental architecture of our culture.

Whatever their race and culture, nearly all the people I have known who have experienced these plants have reported getting this similar message: it is our pressing responsibility to take care of the biodiversity and the health of this planet and that the plant spirits are enlisting our support. It is no accident that they're producing compounds that stimulate consciousness: it's the way they speak to us and through us.

I only ingest the sacred mushrooms very sparingly now. I don't do them very often, and frankly, I don't think most people should ingest these substances. I don't think everybody should ingest psychoactive plants. Many people are not mature or centered enough. I think the plant spirits have a lot to teach willing students and serious people who are ready to listen, but these plants have been dangerously commercialized and popularized, and that can pose real problems.

Mushrooms and these other plant medicines make you face your soul. Are you ready to look at your soul? Are you ready to look at who you really are, including your deepest fears and failings? Some of us are not ready. I know lots of people who are not ready. They don't really want to deal with themselves, and the plant spirits and the mushroom spirits are very good at putting a mirror up to your face and making you deal with deep issues and issues way beyond your own ego. And so as I get older, my respect for the plant spirits increase, and my conservativeness about their use also increases. In my youth I was on a mission to convince everybody to use these things, then I realized over time that they are only appropriate for certain people.

Kat Harrison: I agree. In my fieldwork of over twenty years in the Peruvian and Ecuadorian Amazon, Central America, and Mexico, particularly among the Mazatec people in Oaxaca, I have observed that all of these powerful plant teachers are only used in a context of prayer. And prayer always begins with gratitude and saying thank you for what we have. It then proceeds through varying forms, and then it ends with gratitude again. That's really the model for having a relationship with these powerful plants, whether you're doing it ceremonially in a tradition or in your own form of ceremony. There really is no appropriate casual use of these teachers.

Using any plant for medicine or even for food or ornamentally puts us into relationship with it. It's an invitation for an intimate connection with it, for it to come into our lives as a being. Some of us, for example, surround ourselves with plants that bring a certain quality that we hope to propagate, literally and figuratively, in our lives. It's important that we not let ourselves fall into unconsciousness in our relationship to plants. A conscious relationship is always its own reward, and we inevitably pay a price for unconscious relationships.

Where the great, thorny, complicated mystery of addiction comes in is when we set up a relationship to a plant unconsciously, without self-awareness and without respect for the plant. Then it feels dishonored. Tobacco addiction is an obvious example of the price of an unconscious relationship. It can make you crave it but never lets you feel satisfied. That's the state many of us are in, and not only with plants.

Dale Pendell: I'm not convinced that being consciously aware makes one immune. We're interested in these plant allies because they have power or healing qualities. They have some virtue or force that we want and, like political allies, some of them may have their own programs. I'm not sure we really understand the plant world, what all of them are up to. I'd argue that even conscious use doesn't necessarily make the path safe. Also, ingestion of a plant is only one way of relating to it. There are whole shamanic paths in which the relation to the plant spirit is achieved through offerings to the plant itself, especially among agricultural peoples for whom specific plants provide the economic basis for the community.

One could argue the greatest hallucinogenic plants are rice, maize, and wheat. They have this effect of triggering great pleasurable sensations and warmth, and, by keeping us alive, they fuel this fantastic hallucination in which things such as electrons and molecules in the air appear to our sight as recognizable objects.

Kat Harrison: Yes, enthusiasts of psychoactive plants tend to focus only on a few species with obvious power, but there are remarkable plants all around us. For example, I have a wonderful relationship with mugwort, specifically *Artemisia douglasiana*. This species of mugwort grows abundantly all along the trails and hills of northern California and many other places. Mugwort and the entire *Artemisia* genus are great medicines. Just touching the oil to your fingers or picking a few leaves (of course, after asking it and thanking it) and rubbing them on your temples is very clarifying.

It's also renowned as a dream medicine when you put some under your pillow, and I find it a great smoke. I like to smoke, and I like to dry leaves or flowers of plants that have caught my attention (and that I know enough about to know they're not toxic) and to smoke a tiny bit. I find it helps me recognize the qualities of that plant immediately. I find that mugwort leaf smoke really clears and refines my mind and doesn't have any distorting properties.

Dale Pendell: I would add that it's also a superior substitute for what they call the small absinthe or the small wormwood in absinthe. Instead of using *Artemisia pontica* you can use mugwort.

Kat Harrison: It's also important to realize that not every plant one of us or anyone else recommends is going to be every person's ally. Like with people, some will be your friends, some just casual acquaintances, and some won't be your cup of tea, so always start with a light touch, build your relationship slowly and pay attention. Some plants may be allies of yours, but are too tricky to share with others or to speak about. Also, there may be plants that are best left alone. Some are very fragile and endangered, and their commodification and desacralization threatens their survival. Some may only be meant for people from a specific group. Some may not be appropriate for humans.

Dale Pendell: I've noticed that some plants tend to stay hidden. It's part of their history and nature. If you have a relationship with such plants, I think you should be very careful about with whom you share them. I'm writing a book about some of these plants, and I've discovered it's very difficult to decide the appropriateness of discussing or sharing potentially dangerous knowledge, especially because our culture lacks a living tradition of this type of wisdom. I think the best tack is to seek to learn from those who have such traditions.

Kat Harrison: And it's even more crucially important for us to be mindful because a lot of this wisdom that comes from plants and similar enlightening agents is perceived to be very threatening in our culture.

Paul Stamets: Yes, that's right, and it's a shame because some of these plants could have highly beneficial uses. The situation is improving slowly. Formal, legal research on psychedelics is starting up again around the world after decades of inactivity. There have been conferences on the medicinal properties of psilocybin in Japan, for example, that featured papers on promising treatments for certain types of schizophrenia with psilocybin. Even in the United States there are now signs of growing hope for more medical research soon.

Kat Harrison: I'd like to go more deeply into our relationships with all living things and the natural world. I think the type of attitude toward plants that we've been describing, one that respects the intelligence of other species, can take us toward very different ways of relating to all life forms. For example, some micro-flora and micro-fauna cause disease in many, many people. A year ago I contracted hepatitis A while I was doing fieldwork in Mexico. I didn't realize it until I got back, and I spent several months pretty wiped out while I tried to carry on my complicated day-to-day life. I really just had to take it easy for quite a while.

But I reminded myself that this was an organism with an identity of some sort and that therefore I should meet it person to person. When I did so, I would lie there under an invisible shroud of sort of ugly yellow that colored everything I looked at. I felt as though I were lying under an old car looking at my own broken engine for a couple of months. I could see everything that was wrong, but I couldn't do anything about it. I did attempt to ask this virus living in my body: "Who are you? How can I understand you? While you're here, while we're sharing the same space, can I get to know you and understand something about you?"

That helped give me a sense of what I could take to bring my own system back to equilibrium and back to health. I found an interesting story from the sixteenth or seventeenth century about the period during which smallpox reached a certain area of the Peruvian Amazon, after contact with the European colonizers. The disease came over the Andes and into the Upper Amazon, where a large tribal area was being ravaged by smallpox.

This undoubtedly happened in many places, but this particular instance was actually recorded in various surviving written accounts. Apparently the medicine people of the tribe went into council with their plant allies and said: "We need to understand the spirit of this disease which has come to meet us." They apparently came to perceive smallpox as a being with certain personality traits, and they decided on ways of interacting with it in order to protect themselves from it.

One of the things they did was put all of their clothes and jewelry on backward, so they always appeared to be walking away from it. There wasn't enough documentation to tell how this all came out. We know, of course, that smallpox took a terrible toll, an appalling toll, on the peoples of the Americas, and the odds are strong this backward dressing strategy didn't work. I'm certainly not advocating that we dispense with modern medicines to cope with dangerous diseases, but it's the approach that this anecdote highlights that interests me, this willingness to ask every new entity we encounter, "Who are you?"

The micro-flora and the micro-fauna of the world are really running the joint, actually, so it's especially important, I think, to include them in our understanding that all levels and forms of life are, in some ways, conscious and in relationship.

AUDIENCE QUESTION AND PANELISTS' RESPONSES

Audience Member: I have the honor to have a lot of young people in my life. Do you have any advice on how to relate to the use of sacred plants among young people?

Jane Straight: I think the best thing those of us who do have sacred plant use in our lives can do for our children is to set an example of respect, moderation, mindfulness, ceremony, and conscious intention in our own relationships to these plants.

Paul Stamets: Three of us here have teenage children, so this is a subject that we have personally talked about on more than one occasion. It's a tricky issue. One thing we all acknowledge is that we've lost the art of ritual in our culture, and without the structure of ritual it's really hard to bring younger people into this knowledge in such a way that they can respect it the way we've learned to respect it. Speaking to teenagers is rarely easy. Often when you try to talk to them their eyes just glaze over. Unfortunately in most cases they need to discover things on their own, and the road of life is full of pitfalls and tragedies and painful experiences, and perhaps that's what all of us need to go through.

Unfortunately, with the average American watching four to six hours of television per day, our culture certainly doesn't provide fertile ground for healthy ritualistic structures for the introduction of sacramental plants. I think killing your TV, and setting up a temple and making an altar out of it is a really good place to start.

Dale Pendell: It would certainly help if the use of sacramental plants were already integrated into the fabric of a community, as it is in many indigenous cultures. Of course, the legal status of these plants in our culture makes that very, very hard to do, so I don't have any solution to that problem of youth, but I do have a story:

Long ago there was a god who would appear from time to time, a young god associated with wildness and wild places and associated with ecstasy. His name was Dionysus. It was recognized that he was intimately connected with growth and spring and all of the things that make life happen. There was also a kind of wild, uncontrolled aspect to his appearances—clearly a dark side to it. No one would call Dionysus safe, but the worst thing you could do was to deny this god or to fail to recognize him. The way the myth goes, the penalty for not recognizing Dionysus was the sacrifice of your own children.

This discussion took place at the Bioneers Conference in 1997.

12

A North American Indigenous Look at Sacred Plant Use

Katsi Cook

IN MEMORY OF OUR BROTHER GARY RHINE

I was born and raised on a Mohawk reservation on the St. Lawrence River, right on the U.S.-Canadian border. It is trisected by three governmental jurisdictions: Quebec, Ontario, and New York State, so we've had many complexities to deal with in our development as a people and as a community. As part of that development, in my own life, I've committed to that area of our sovereignty that has to do with control of health and reproduction. My generation has focused on the protection of indigenous knowledge and other areas of our sovereignty, such as the control of our land base, control of our education, control of our psycho-spiritual life and the ability to solve disputes among our own people. In protecting our medicines, for example, we must first never forget how to be in relationship with them. This is the first protection. Talking about them is one thing; using them to support life, such as the life of one's own child or grandchild, is quite another. I therefore base my work and my teachings in an ecological approach, one that considers networks, balance, cycle, and flow and the patterns and processes by which nature continuously creates and sustains life.

Our Mohawk creation story is a study in sustainability. It is the story of a pregnant woman who falls from the sky, transforming spirit into being human on this Earth. In fulfillment of the dream of the Chief of the Sky World (who dreams that the fruits and flowers of a great celestial tree that grows at the base of the Sky World are wilting and dying), Sky Woman, who has come into relationship with him, uproots the tree as she pulls at a flower that grows at its base, thus creating a hole in the floor of the Sky World. Leaning deeply into the hole to search the great blue sky below, she falls through.

As she falls, she grasps at the edges of the hole, trying to pull herself back up, but she does fall, and with seeds and bits of sacred things from the uprooted celestial tree now embedded under her fingernails. Grasped in her hands are the original tobacco, the strawberry, and other medicines that we now depend on here in this world. She lands on Turtle's back with the help of the winged ones who guided her fall to Earth. Sky Woman's name—*iotsi tsisoh*—means "mature flowers." We can think of her name as a name for some of the medicines we use, and we can refer to them collectively as *ononkwa sohna*—"medicine people."

Establishing herself on Turtle's back with the aid of the water animals (who brought to the surface from the depths of the ocean a handful of dirt in which could be planted the seeds she carried), Sky Woman soon gave birth to a daughter. In the company of her mother, the daughter quickly grew and, when she came of age, she had a number of suitors who wanted to marry her but her mother refused to give her permission. However, one handsome young being (who turned out to be Turtle) impregnated her and she bore twin sons into this world, who, in their dialectical relationship, their arguments with each other about the way this world should be organized, became the Creator Twins who created the world as we know it.

Our creation story teaches us that there is no complete good or complete evil, and that in order for this world to exist, there has to be a balance. We live in a universe of relatives, and the universe is kept alive by those relationships. I have heard people define ecology as a science of relationships, and that's certainly how I perceive my own work—keeping those relationships going to continue the work of our celestial mother.

The strength that I carry comes to me from my mother and my grandmothers. I was delivered at home by my grandmother in 1952, to a mother who was told not even to have children because she fell in the river when she was a little girl and got rheumatic fever and rheumatic heart disease at a time when there were no antibiotics. The rheumatic fever damaged the mitral valve of her heart. She brought me, her fourth child, into this world at risk to her own life. These women that I come from are the ones who taught me at a very young age about the power of belief and the focused intent and concentration of prayer.

The word *midwife,* meaning "with woman" has roots in old English and German. It's a good word, also signifying relationship, but it's limited in scope. In my community, one word for midwife is *iewirokwas,* "she pulls them from the water, from the Earth, or a dark, wet place." This description of what a midwife does is full of ecological context. To begin to integrate our traditional teachings in a real way, we've had to go back to our language and to those things that grandma and our elders held on to for us. To handle medicine means that you respect all of those relationships in that reproductive ecology that is this great universe, this great womb that we are all related to.

One of the things we have tried to do is to support the strength of our traditional medicine societies, so they can do the work that they were given to do for our people. To improve the respect, the equity and the empowerment of that sector of our knowledge base has been one of my goals. In strategizing a community response to the industrial contamination of our local food chain and subsistence economy, I, along with others in my community, started to organize our own knowledge of traditional medicines.

With support from a Partnership in Communications Grant from the National Institute of Environmental Health Sciences' First Environment Communications Project—working with St. Regis Mohawk

Health Services—we put some of our primary care providers through a six-week training period with some of our medicine people.

The National Institutes of Health would categorize this knowledge as "Complementary and Alternative Medicine," but it is simply, but no less significantly, the cultural knowledge we have had to depend on to maintain our health and well-being as Mohawk people through the generations. Our resilience and survival is based on it. So, of course, these things deserve their proper respect. To use these traditional medicines, you must first know how to respect yourself.

We still follow those old ways of collection, of prayer, of relationship, that go with using these medicines. As a midwife, when I ask a mother to begin to use, say, slippery elm two weeks before her baby is due, there's a whole protocol to be followed. I have to find an elder in our community who's going to fix that medicine for her, because it isn't just about stripping the inner bark from a live tree. It involves the right approach to the tree, the proper respect and address to the tree. For instance, in addressing a plant, you must use your Mohawk name, gained through birth into a clan such as Turtle, Bear, or Wolf, and given to you in ceremony to establish the proper relationship of each individual child to the natural world, including the medicine plants.

The road to know about plants is long. In my communities, you usually can't even ask directly about plant use. It's disrespectful. Most people come to the plants because of a need: they're hungry, or sick and maybe even dying, or they have a need in terms of protection. So the plant knowledge comes to one such as myself in titrated doses, and it's not possible to know everything about every plant. Just like at the clinic, there are specialists and there are generalists.

Every indigenous culture has its body of plant knowledge. It is never the plants alone that heal. Healing is a process that involves context and connection. Knowledge and healing come through dreams, through birth, through ceremonies, and through the practice of private tradition and family ways. This information moves along through family lines, so that certain families hold certain medicines, and they are respected for the spiritual ways that they must fulfill so that they can continue to practice as healers.

Not all knowledge is for everyone. That's just the way it is. There are those in our community, usually women past the time of menopause, who are the ones to go to when you need a medicine. There are men too, of course, but in indigenous societies the healers are mostly women. I think this is the natural extension of our knowledge about menstruation and birth—both powerful aspects of the human experience. I'll often send the mothers under my care to these different healers for help, and they're quite capable people, but they won't always share their knowledge. They'll say, "We keep our knowledge close to us because that's something we've learned after a life of struggle and trial and commitment—it's not for everyone."

Our people believe that you can't communicate with the spirits of these plants unless you have an Indian name, and for that purpose, when our babies are named in our long house, there's only one of that name given, so it won't confuse the universe. The clan mothers hold these names in a metaphorical bag; they are like property. You can't just pick up any old name and use it. The Indian Health Service spent twenty years getting our people away from our plants and traditions, and in their place offer expensive pharmaceuticals that, because of poor nutrition and low activity levels, many now depend on to sustain their lives. A lot of people lost touch with this plant knowledge and, to such an extent that, for example, young mothers no longer know the difference between a sick baby and a baby they can care for at home themselves.

I have a friend who is a research physician who, in the course of his research, fed lab rats the congeners of PCBs most commonly found in Mohawk mothers' milk in order to identify effects on the rat litters. For about six months he walked around with a respiratory ailment; he couldn't stop coughing. You couldn't make him laugh at meetings because he'd begin a coughing fit and couldn't stop. Due to the fact that he works with lab animals, when he started coughing up blood he was quarantined and tested for both the animal and the human forms of TB.

While he was in quarantine, I sent him some sweet flag (*Calamus acorus*) and told him, "It's not like the medicine you're used to. You have to respect it in a certain way," and I told him how to use it, explaining that you have to develop a relationship with a medicine, just like a relationship with a friend. If it is a really powerful medicine, then you might even address it as "Grandfather" or "Grandmother." You are supposed to talk to the medicine like you're talking to a friend. You're supposed to take the time to be with the medicine. And then after a time when you start to feel better, you are supposed to put the medicine back onto the Earth, with gratitude and thanks, in a place where people aren't likely to go.

About two weeks later my friend called me and asked, "What is that stuff you gave me?" For the first time in six months he had stopped coughing. I was taught by my grandmother that sweet flag is used for anything in the head or the chest, from a toothache or headache to a sour stomach or chest congestion. She used to send it to me and to my sister at boarding school, and the Catholic nuns would smell its pungent odor and ask us what it was. We never did tell them. We'd tuck a knuckle of the root in our cheek in the winter when we got chest colds or sore throats. We would walk around with it in our mouths until it swelled up, soaking up the toxins from whatever was ailing us.

The sweet flag lives in the wetlands, a special environment indeed. It is accustomed to being where there is a lot of moisture and fluids, so when it is picked and then dried until it shrinks to about a third of its size when fresh, it then absorbs fluids from the mucus membranes of the mouth, and its healing spirit passes into our bodies through those same membranous tissues. I told my friend what it was and how useful it was and how all my life I'd been using it, even as a little girl.

He said he was really happy to use a medicine that was a root that came out of the Earth—a whole different idea than taking a pill. Among the nice things he had to say about sweet flag, he commented that it improved his mood. I'd never heard that it could improve your mood, so I went to Jim Duke's database of medicinal herbs and I read that beta-asarone is the chemical constituent responsible for the "mildly hallucinogenic" properties of this plant. I went back to some of the elders—including the one on whose property I'd collect the sweet flag with my sister every fall—and asked them about my friend's observation. They said that that's part of its healing power.

Somehow the Creation knows that when you're not well it affects the way that you think about yourself. If you've ever known serious illness you know how it can make you feel depressed. If you have to depend on others to feed you, dress you, or you can't move around like you're used to, you can get pretty low in spirit.

Medicines are not just plants; they are also the practice of dreams and ceremonies that go with them. We have a relationship with a Mayan elder in Guatemala who my husband and I have known since 1981. In learning aboriginal midwifery, my focus is on interpreting and integrating indigenous knowledge with the biomedical skills necessary to be a safe practitioner. Our indigenous knowledge has not been lost. It evolves like everything that is alive, and it is recovered in the practice of doing, of using it, of believing in our medicine. So the medicines I use in my life as a mother, grandmother and midwife come from the four directions. From the south I study the ways of the days of the Mayan people, the Mayan ways of the days that are still used among their midwives and day-keepers.

In my journey to become a midwife I had gone to the Farm in the fall of 1977 to study with Ina May Gaskin and the Farm midwives who are in the book, *Spiritual Midwifery*.* With the help of my friend Ann Boyer, an OB-GYN, I attended the University of New Mexico Women's Health Training Program for my clinical training the following year.

---

*The Farm is an "intentional community" in Tennessee that was founded in the early 1970s under the leadership of Stephen Gaskin. One of the most successful, enduring institutions to come out of the counterculture of the 1960s, it became world-famous because of the pioneering work done by Ina May Gaskin at its Midwifery Center.

I'd see so many native women in the clinics there—Zuni, Pueblo, Navajo, Hopi—who had classical caesarian scars, and they had no clue why they had had a caesarian and I thought, *This isn't right. We have such beautiful teachings, but we haven't integrated the spirit of these teachings into our physical experience of being a woman.* So, after my clinical training, I went to stay with my sister-in-law, Loretta Afraid of Bear Cook's people on the Pine Ridge Reservation in South Dakota. Her mother, Beatrice Weasel Bear, was a midwife. I wanted to learn *traditional* midwifery, not just the biomedical model.

Beatrice said to me: "Daughter-in-law, if you want to learn midwifery, you have to go in that tepee." For over a hundred years the Native American Church of South Dakota has been holding the peyote medicine in what's referred to as the Half Moon fireplace. The ceremony Beatrice introduced me to was a "going back to school meeting," a ceremony of the Half Moon fireplace for which Beatrice and Loretta serve as water women, or women who carry water from which ceremonial participants receive spiritual help.*

I said, "Where I come from, in the longhouse, we are told not to use any mind-changers." My sister-in-law looked at me and said, "Well, I don't think you know what this medicine is. This medicine is the heart of the Creator. We call him Grandfather."

Loretta told me a story about her mother who had gone into labor—with Loretta's younger brother Aloysius Weasel Bear—while attending a peyote meeting. The road man made peyote tea for her to help her in her labor.† It was a powerful story that I never forgot, a story of the efficacy and strength of medicine. And so I went in that tepee. My sister-in-law

---

*Beatrice and Loretta are both featured in the video *The Peyote Road,* the 1994 documentary that explores the historic legal struggle undertaken by the Native American Church of North America (NACNA) to support the religious freedoms of Native Americans. The NACNA was founded by the last chief of the Comanches, Quanah Parker, in the late 1800s.

†The "road man" is the designated leader of a peyote ceremony in the Native American Church.

promised me that whatever I wanted to know, this medicine would show it to me. I soon realized that the ceremony itself, which begins at nightfall and continues through until sunrise (twelve hours) is about birth, involving similar energies and sacrifice.

It was a back-to-school meeting after all, and I was seeking natural knowledge, so during the night I asked Grandfather-Medicine, who would become a great teacher to me: "How do I know I can be a midwife? That's a big responsibility. How will I know what to do?"

Grandfather spoke clearly, "You'll know. Just do it!" (This was about twenty-seven years ago, way before the Nike ad came out.) Ever since then I've been using the medicine in my own life, including to help give birth. I used the peyote medicine to help me go into labor with a set of twins at the age of forty-one. It worked! The medicine helped me to pay closer attention to my body and the messages of my unborn—messages which arrived in a cascade of reproductive hormones. I have never had an induced labor or had to use sedatives or painkillers (which are notorious for their negative effects on the newborn).

More recently I was also lucky to be able to spend time with and learn more with Grandma Guadalupe de la Cruz, a Huichol midwife and healer, before she traveled on ahead of us to the spirit world. You have to be a card-carrying Native American, a tribally recognized enrolled member, to legally use *Lophophora williamsii* in this country under the aegis of the Native American Church of North America. Several years ago a nurse from an Indian Health Service (IHS) hospital in the Southwest came to the NACNA annual organizational meeting and asked the national organization to send someone to her hospital to help train the caregivers there. Apparently some of the nurses at the hospital were calling the Drug and Alcohol Services whenever they found a native woman using peyote in childbirth. This is the kind of cultural denigration that Native American people still have to endure for something we relate to as the "heart of the Creator." Peyote can heal anything, and when I say that, I don't mean it's a panacea or a cure-all. I mean that it can restore people to a place where they can allow the homeostatic work of their own body's healing by getting themselves out of their own way.

Plants are like people. You can only have so many relations that you have time to nurture. Some people are even running to the rainforest in Peru to find medicines, and I do have to admit I've been to the Mayan people of Belize, and they do have beautiful medicines, but it's the knowledge itself that is strong and powerful in the use of the green medicines. Not everyone can use all of these medicines. Peyote, for example, doesn't belong to everyone. I ask you in a humble way to protect the exemption in the American Indian Religious Freedom Act.*

We are looking for a niche in our worlds to continue to survive, and this medicine has come to us in a very organic way, not through the forces of markets and economics. We take that very seriously. The other medicine I use is, of course, the medicine of my own people. There are groups of medicines that like one another, that work together. In doing our ecological analysis of how toxins move through our environment, we were shocked, for example to find that at Akwesasne, along the St. Lawrence River, there's an increased uptake of methyl mercury by plants in the wetlands areas. This was especially true of another one of our very powerful medicines, the yellow pond lily.

That medicine is supposed to be picked in a certain way, usually by young people who have to keep their mind a certain way when they gather it. The root grows way deep into the muck of the wet areas, and if your mind's not strong and your heart's not pure about the person that you're trying to help, I've heard that the roots can pull you under

---

*The American Indian Religious Freedom Act, passed by Congress in 1978 and amended several times since, allows, among many other provisions, the use of peyote for legitimate Native American religious use by Native Americans, thereby creating an exemption in the drug laws prohibiting peyote use, but only for this purpose. One of its passages reads: "Notwithstanding any other provision of law, the use, possession, or transportation of peyote by an Indian for bona fide traditional ceremonial purposes in connection with the practice of a traditional Indian religion is lawful, and shall not be prohibited by the United States or any State. No Indian shall be penalized or discriminated against on the basis of such use, possession or transportation . . ."

and take your life instead of helping you to assist with another life. Mohawk stories can get pretty scary. The stories make you think about what you're doing.

At the same time, I know that the availability, the access to our own medicines, is diminishing—given the increasing encroachment of development and the loss of land-base to toxic contamination. We are looking to land claims and land restoration to try to solve the problems of access to our own medicines, but we also need to be propagating and growing these medicines. Part of my training as an aboriginal midwife was to raise, for four years, a field of corn from seed that came with Sky Woman in her fall, taking care of it, singing to it, dancing with it.

The Mayans teach that women learned midwifery from corn, and the kind of corn we grow can't be harvested by machine. It has to be hand-collected, and we'd have about twenty people out in our fields to harvest this corn. When you pull back the husk, it's like seeing a new baby. You might make a comment like: "Oh, look at this one red kernel among all the white kernels. It must be related to the Tuscarora corn."

When I had the wife of one of our Wolf Clan chiefs come to visit my home, I was telling her about this, and she said, "Well you know, the word in Mohawk that describes bundling a new baby means, in English, 'she's putting the husk back on the corn.'" Every part of corn is a medicine, including the silks, which are used for the urinary tract. A soil scientist, a Ph.D. agronomist at Cornell University from Nicaragua told me how the gestation of the corn and the gestation of the human being are very similar, so that in that way, corn teaches us about midwifery, about genetics, about the secrets of the plants.

I find that whether we're talking about indigenous knowledge or biomedical knowledge, it's the same truth. For example, one of the strongest teachings we have is, "believe in your medicine." Efficacy is all tied up in physical and psychological processes. From testing with experimental drugs we know the profound power of patient expectation (the placebo effect) on the individualized ability of the brain to release natural painkillers, our endogenous medicines.

Indigenous medicines and their uses must be restored to our people so that we can again be healthy. In my community we have very limited healthcare dollars, and environmental health impacts are going to drive up the prevalence and incidence of diseases that have an etiology in the immune system. A lot of the medicines we used were protective of health. We need to restore those behaviors, those attitudes and those practices that five hundred years of colonial oppression have drummed out of us in our communities; they have been kept underground and need to be supported so they can flourish. We have to find ways in our communities to make a place for the medicines to sit with us once again.

This presentation took place at the Bioneers Conference in 1999.

PART FOUR

Brazil's Modern, Entheogen-based Religions

13

The New, Syncretic, Ayahuasca-based Religions

Luis Eduardo Luna, Ph.D.

I've been studying ayahuasca since 1971 from different points of view. As an anthropologist, I have been interested in the role of ayahuasca in the medico-religious worldview of the people who consume it, in the process of knowledge acquisition through the brew, and in the artistic manifestations of the ayahuasca experiences as well. I have also collaborated with various scientists in the study of physiological correlates of the ayahuasca experience.

The ayahuasca brew is made by combining the crushed stem of a *Malphigiaceae* vine (*Banisteriopsis caapi*) and either the leaves of *Diplopterys cabrerana* (a vine of the same family) or the leaves of *Psychotria viridis*, a small tree of the *Rubiaceae* family. It has been used by many Indian tribes of the Upper Amazon, probably for millennia. It permeates the religious and artistic lives of some of these indigenous groups. The art of the Shipibo and Tukano peoples, for example, is completely inspired by the visions elicited by ayahuasca. The narratives of the Siona Indians of Colombia, according to anthropologist Jean Langdon, deal almost entirely with their shamans' adventures on ayahuasca journeys.

Ayahuasca use eventually spread beyond the indigenous world. Mestizo jungle/river-dwelling *vegetalistas* or *ayahuasqueros* in Peru, Ecuador, and Colombia adopted the use of this brew from the Indians— to heal, to get in contact with the supernatural world, and to perform what could be called shamanic deeds. Between 1980 and 1989 I studied this tradition; it was the topic of my Ph.D. thesis. Most of these mixed-blood practitioners speak only Spanish but have ancestors from indigenous groups from the Amazon and also the Andes. They view ayahuasca as a plant teacher. They call it a "doctor."

Several other species, some of them used as ayahuasca admixture plants, are also considered plant teachers. For the mestizos (as is also the case among the Yagua Indians, according to Jean-Pierre Chaumeil—a highly respected French ethnographer/anthropologist who has done extensive field work in the Amazon) the plants are considered to be the real path of knowledge; it is believed that they teach medicine. Illness is usually conceived of as being caused by an animated agent, and to involve psycho-spiritual factors as much or more than physical ones.

In order for the practitioners to learn their art, they take the plants while keeping to a very strict diet in isolation during a period of time, sometimes three months, sometimes six months, sometimes even years. The spirits of the plants appear to them, either in visions or in dreams, and teach them this form of psycho-spiritual medicine.

In 1985, through Dennis McKenna, I met a former *ayahuasquero* named Pablo Amaringo. Over many years, we established a very intimate relationship. Among other talents he was a skilled landscape painter, although he did not make his living with his art. I suggested to him that he paint the visions he had had when he was a practicing *ayahuasquero*. The result was a great surprise to me, because it was like being able to penetrate visually the cosmology of the *vegetalista* tradition. I could, for example, see clearly how the shamans conceive the jungle as not only filled with plants and animals but also as full of spirits, full of intelligence. Shamans use the plant teachers to connect with these spirits, which are the repositories of the wisdom of the forest.

Out of our collaboration came the book *Ayahuasca Visions: The Religious Iconography of a Peruvian Shaman* in which forty-nine of Amaringo's paintings are published, together with his descriptions of them and my notes relating ideas and iconographical motifs found within the Amazonian and Andean worldview.

A very important element in Amazonian shamanism is the use of songs inspired by the spirits; mestizo practitioners call them *icaros*. In 1986 I spent a month with Don Basilio Gordon, a Shipibo shaman living in Santa Rosa de Pirococha, a small settlement on the shores of the Ucayali River in Peru. I was interested in which other plants, beside those involved in the preparation of ayahuasca, he knew about. He was treating patients, but I never saw him use any plants. I asked him about it, and he said that when, through ayahuasca, one gets in touch with the plant spirits, they will teach you their songs, and you can use these songs to heal. Only if the song doesn't work do you have to go back to using the actual physical plant.

Once Pablo Amaringo improved his living conditions through his art, young people went to his home to learn to paint. In 1988 I proposed to him that we create a formal school of painting. It was called the Usko-Ayar Amazonian School of Painting. The idea was that through art the students would learn about their own environment. I discovered that, when given the opportunity and proper training, Amazonian young people are very artistically gifted. They seem to have an excellent eidetic memory.

Pablo Amaringo taught these art students how to observe nature carefully, how to keep precise images in their minds, and how to create—in their minds—a landscape with those elements, respecting "reality" in terms of the accuracy of the depiction of natural objects (plants, animals, rivers, and people) and their relationships (ecology). It became a very successful project. The school attracted a wide range of people, including Elías Silva, a Shipibo artist who illustrated Shipibo myths in vignettes.

During its peak year, in 1994, it had about three hundred students who paid no fee and who received all art materials free of charge. When there were sales, the students would receive 50 percent of the benefits of their paintings while the other 50 percent would go to the school to meet its needs. We had exhibitions in several European and Latin American countries, the United States, and Japan. Most of the work was by children between fourteen and eighteen years of age.

I worked intensively on this project until the beginning of 1995. Then it became too heavy a burden on my shoulders, as new academic duties called me to Brazil. I made the hard decision to leave the school. Sadly, it seems that it became very difficult to administer, and I gradually lost contact with Pablo Amaringo. I understand that he still receives students from time to time in his home in Pucallpa, but the school as an institution is no more.

However, I learned with great joy that some former students became art teachers in Peruvian schools while others had exhibitions in various countries and some even founded art galleries. Others became forest guides. The artistic career of Anderson Debernardi, perhaps Pablo Amaringo's most gifted student, has been the most interesting. Debernardi made huge murals for the Helsinki Zoo, illustrated the books of famous ornithologists, and has had exhibitions in several countries.

In July 2005 I was invited to speak in Iquitos, Peru, at a conference on Amazonian shamanism. The hotel in which I was staying was thoroughly decorated with paintings in the typical style of the Usko-Ayar School of Painting. I am glad to see that the project's impact has continued through the lives and the work of those who had the great fortune of being touched by it.

Back in 1984, Clodomir Monteiro da Silva, a Brazilian anthropologist, invited me to Rio Branco, the capital of Acre State in the Brazilian Amazon. There I had my first contact with a new phenomenon, the use of ayahuasca as a sacrament—incorporated in the practice of three religious organizations. In 1991 one of them, the União do Vegetal (UDV) invited me to their first scientific conference, and I was able to realize the importance of ayahuasca in Brazil. In 1995 I began teaching anthropology at the Universidade Federal de Santa Catarina, in Florianópolis in southern Brazil, and between 1995 and 1998 I visited the centers of the three main such religious organizations in the Amazon area and in other regions of Brazil.

This Brazilian phenomenon has its origins in the 1930s, when a humble black rubber-tapper called Raimundo Irineu Serra (1982–1971) met a Peruvian shaman and took ayahuasca for the first time. It is said that in his visions he saw the Virgin Mary as the queen of the forest. She told him to start a new religion using the brew. In Rio Branco he created a community and gradually developed a doctrine combining popular Catholicism, Kardecism (a religion inspired by the writings of Leon Denizarth Hippolyte Rivail), and other Afro-Brazilian and European esoteric ideas.*

The original group splintered after Mestre Irineu's death. One of the branches—called CEFLURIS (Centro Eclético de Fluente Luz Universal Raimundo Irineu Serra), also known as Santo Daime (the name they give to the sacred brew)— spread all over Brazil and eventually was also established in several European and Latin American countries and in the United States and Japan. Today CEFLURIS has about twelve thousand members; they associate the *Banisteriopsis* vine with strength (*a força*) and the masculine, with Jesus, and with the Sun, while the *Psychotria* plant is associated with light (*a luz*), the feminine, the Moon, and the Virgin Mary. In their worldview, the two plants come together to sym-

---

*Leon Denizarth Hippolyte Rivail, better known by his *nom de plume* of Allan Kardec, was a nineteenth-century French author and esotericist whose writings about Spiritism had a wide influence, especially in the "New World."

biotically harmonize these two polar energies and give one revelatory, benefic visions. A consequence of these ideas is that the role of male and female is well-balanced.

A second religious organization, which uses ayahuasca as a sacrament, is the União do Vegetal (UDV), which was created in 1961 in Porto Velho, the capital of Rondonia in the Brazilian Amazon, by Jose Gabriel da Costa (1922–1971). Mestre Gabriel (as he is called by his followers) was born in Bahia, in the Brazilian northeast, but emigrated to the Amazon to work as a rubber tapper. The name "União do Vegetal" is interpreted both as the union of the two plants and the higher sense of union experienced through the use of the ayahuasca brew, which they call *vegetal*. Starting from humble beginnings, like the organization created by Irineu Serra, the UDV spread first to Manaus, then to Sao Paulo and the rest of Brazil. It now has over sixteen thousand members, including lawyers, doctors, academics, people in government, and all sorts of other professionals.

Santo Daimé and the União de Vegetal have a different ethos. Santo Daime's iconography is quite flowery and elaborate, while the UDV's is much more discrete, with very little imagery. The União do Vegetal seeks to permeate society, to gradually change society for the better, whereas Santo Daime offers its members, in many instances, a retreat from society. Their ideal is the creation of a community in the forest; to live in nature.

The forest is also present in the UDV, but in a different way. In order to become a *mestre,* the highest rank in the four-leveled hierarchy of the UDV (to which only males have access), one has to know the plants in the forest and how to prepare the hoasca brew. A real assimilation of the Santo Daime religion implies spending some time in the forest, whereas to a certain extent the philosophy of the União do Vegetal is to bring the wisdom of the forest to the city.

Every center tries to grow its own plants for the brew, so that the local people have an intimate connection with the plants that make their sacrament. União do Vegetal members are largely urban and are, in general, environmentally conscious; because their sacramental plants grow in the forests, they advocate forest conservation.

The UDV is hierarchical but also democratic, since all positions are subject to election. Elected members need to demonstrate by their behavior that they deserve their respective position, which can be lost at any time. Power tends to stay in the hands of the wisest members, and the organization is structured in such a way so as to prevent the accumulation of excessive power in any one individual. The UDV is an initiatory religion with an oral tradition. It stresses the importance of memorizing certain *histórias* (stories) and *chamadas* (songs), something that brings to mind indigenous traditions such as that of the Tukano in the Colombian Amazon.

During my research I got in touch with a third religious organization, which had until then not been mentioned in the scientific literature on ayahuasca. It is called *Barquinha,* "the little boat." It was created in the forties by Daniel Pereira de Matos (1904–1958), a black man from the Brazilian northeast. Pereira de Matos met Irineu Serra, the founder of Santo Daime, and took ayahuasca with him, apparently several times. He once took ayahuasca by himself and had a revelation. He saw an angel bringing him a blue book with a doctrine, which he then developed with the consent of Irineu Serra.

Pereira de Matos was the son of a slave and little is known about his life. He came to the Amazon in the 1930s, after having been somehow connected to the Brazilian navy since he was seven. Perhaps this is the reason why the church is called Barquinha (the little boat), and also the reason that many of its symbols are related to the sea. He was a musician, made guitars and violins, was a barber, and worked in various professions. He became a great healer and created a center in Rio Branco, which after his death split into three independent but interrelated groups.

The followers of Barquinha view their prime mission as charity but, in their view, charity is not only for the living but also for the spirits of the dead; one can assist these spirits to evolve and "reach the light."

Saint Francisco de Assisi is their patron, and the songs and hymns they "receive" during their rituals with the sacred brew are very reminiscent of the magic songs of Amazonian shamans. In this regard, Santo Daime and Barquinha are similar. Members of these two organizations know hundreds of hymns, and the repertoire grows as new members are added.

I studied for some time one of the three Barquinhas, the one created by Francisca Campos do Nascimento, who was in the main church for about thirty-four years and then decided to leave and create her own church, which nowadays has about one hundred members. It is almost a small monastic order, with almost total dedication. In 1992 I created a calendar with the rituals that a regular member would attend. It came to almost two hundred rituals in one year.

The Virgin Mary is important in the cosmology of the Barquinha, but as I participated in their rituals I realized that what they are dealing with is more like a divine female archetype that is much more inclusive than the traditional Catholic figure. Besides Jesus, Mary, and various saints, they have (either in the altar, or in the *conga*—the space where mediums "incorporate" various spirits) images of *Iemanjá*, one of the *Orixás*, the nature spirit/deities of the Afro-Brazilian tradition. *Iemanjá*, the Queen of the Sea, is often represented as a mermaid.

This strong connection with the sea may seem surprising, in that the sea is four thousand kilometers away, but it resonates quite well with Amazonian mythology. All over the Amazon people believe in magical aquatic cities, in mermaids and dolphin spirits who take the shape of human beings. There are also representations in their iconography of the so-called *pretos velhos* (old blacks), the benevolent spirits of black slaves, as well as those of animal spirits. One of these animal spirits is the swordfish, the aquatic manifestation of a spirit that takes multiple shapes in the astral and Earthly realms.

After four hours of preparatory prayers, hymns, and meditation in the church, in which men and women sit separately on the right and left sides of the temple, the main festivities begin. At this time, the people take ayahuasca again, and they dance in a circle in the *terreiro* (a large space near the temple). During these dances the spirits of the Brazilian *Umbanda*—a homegrown mixture of nineteenth-century French esotericism and African religions—descend and are incorporated by the dancers. When people are under the spell of these entities they move in a different way. They exhibit corporal expressions different from their normal ones. Dancing freely, each person is immersed in his or her own thoughts. When the *Erés* or spirits of children are called upon in certain festivities, people incorporating these spirits may play with toys, eat sweets, and speak like children.

There is an emphasis not only on music but also on aesthetics. Some of the members of Barquinha are artists, and in some cases they have painted their visions, as Pablo Amaringo does. There are many young people among the members of this religion. I calculated that the average age of a follower is about twenty-four years old. Unlike the UDV and Santo Daime, Barquinha churches have been almost restricted to the Amazon, with only a few small centers in other areas. It is a fascinating and highly creative religious movement, which may one day have a broader impact.

All these Brazilian religious organizations that use ayahuasca as a sacrament are against any commercialization of ayahuasca, emphasizing the importance of a proper setting. They have amply demonstrated that they are serious and very responsible in how they use this brew. The UDV has sponsored interesting scientific medical and sociopsychological studies to prove that ayahuasca is perfectly safe and even often beneficial to their members' mental health and social adjustment. An independent psychiatric study of children in UDV households found them as well-adjusted or even more so than the average Brazilian child.

The members of all organizations believe in the therapeutic value of ayahuasca in many situations. They even give it to pregnant women to use before and during childbirth, and the children are baptized with ayahuasca. The evidence suggests that the ritual use of ayahuasca contributes to the well-being of its users, helping them physically, mentally, and emotionally.

I continue to be deeply interested in the potential of ayahuasca. I have created a center in Florianópolis, with the intention of promoting the serious interdisciplinary investigations of ayahuasca, particularly the intersection of art, science, health, and spirituality.

This chapter is an edited version of a presentation given at the Bioneers Conference in 1995; Professor Luna introduced some updates into the text in order that more current information be included.

14

The Extraordinary Case of the United States Versus the União do Vegetal Church

Jeffrey Bronfman

THE BACKGROUND OF THE CASE AGAINST THE U.S. GOVERNMENT

As has been established earlier in this book, the União do Vegetal Church (UDV) is a structured religion protected by Brazilian law, founded by Brazilian rubber trapper Jose Gabriel Da Costa in 1961; it has more than ten thousand adherents in Brazil. Jeffrey Bronfman and others founded an American branch of the UDV in 1993 in Santa Fe, N.M. In 1999 customs agents seized an incoming shipment of hoasca and threatened Bronfman with prosecution under the Controlled Substances Act. Bronfman and several other members of the UDV successfully invoked the Religious Freedom Restoration Act in the 10th Circuit. The 10th Circuit Court identified the UDV as "a syncretic religion of Christian theology and indigenous South American beliefs" and stated: "The sincerity of the faithful is uncontested." The government then petitioned to the Supreme Court for a reversal of the lower court's ruling, citing potential "irreparable harm on international cooperation in combating transnational narcotics trafficking."

The Supreme Court then had to determine whether there was a "compelling governmental interest" in forbidding the use of hoasca tea (a bitter liquid derivative of two Amazonian plants: an Amazonian vine and the leaves of a small tree) which the U.S. Government had asserted should be treated as a domestically and internationally prohibited substance or whether First Amendment rights protect such use in some cases.

Jeffrey Bronfman: Plants have been used as tools for gaining awareness of the spiritual dimension of life for tens of thousands of years. They have played a fundamental part in the religious history of humanity. I am a member of a contemporary church called the União do Vegetal, which began in Brazil and continues this noble tradition of using sacred plants in a religious context. A few years ago the government of the United States took legal action against the religion that I practice by initiating a legal process against me. I was threatened with years in prison if I continued to practice my religious faith.

I and some other church members met with representatives of the Department of Justice to try to help them understand the significance of the sacred tea we use in our religious practice. We brought in a group of experts to give presentations. Among them were scientists, doctors, anthropologists, and members of the UDV. The deeply respected and widely celebrated author Huston Smith—one of the world's foremost experts on the planet's great variety of religious traditions, with a very long lifetime of research and experience—came and explained that the use of sacred plants within the context of spiritual and religious experience was common on every continent. Right here in North America, for example, in northern Mexico, peyote has been used for many hundreds of years. In other parts of Mexico, sacred mushrooms were central to spiritual life.

An anthropologist was included in our group who spoke before the

Justice Department. He explained to them that, in ancient Greece, a plant-based sacred tea was central to the cult that worshipped Demeter at Eleusis; this was one of the most important spiritual rites of antiquity. This practice endured for almost a thousand years and nearly all of the leading figures of Greek civilization took part in it. The idea of drawing intelligent inspiration—in the form of visions from communion with the plant realm—was therefore a totally accepted part of the civilization that our Western tradition is founded upon.

Despite our best efforts, however, we were unable to negotiate any agreement with the Justice Department's representatives that would have accommodated our religious practice. After eighteen months of almost weekly contact with the U.S. Attorney's office we filed a lawsuit in Federal District Court against the U.S. Customs Service, the Department of Justice, and the Drug Enforcement Agency who were all continuing to consider our religious practice a criminal activity. Their position was that our religious use of hoasca violated both domestic and international drug control laws. As a result of initiating this legal action, we had to get very involved in a heightened study of national and international laws on religious freedom and plant medicines and materials.

A key issue was the historical foundation for the notion of religious freedom in United States law. The founding fathers studied systems of social organization around the world to develop a system of government that gave the greatest possibility for human advancement and freedom. They tried to identify those elements that helped societies prosper and those that complicated and inhibited human success and happiness. One of the most important areas of their concern was the relationship between the state and religion. They observed that where people were allowed to freely exercise their religion and approach the magnificence of life without any restrictions or controls by the state, societies tended to prosper.

They decided that the government should make no law establishing religion, that the state should not become involved in sanctioning religion, and that the state should not prohibit, in any way, the free exercise of religion. This is fundamental to the design and architecture of our country. James Madison, who wrote the language of the First Amendment to the Constitution that included free exercise of religion, wrote that when government begins to interfere with the religious freedom of its citizens, it is the beginning of tyranny.

We also did extensive research on the history of plant use for spiritual purposes in this country. We discovered, for example, that at the time of the founding of the country, a tribe of Indians in Delaware was known to drink a tea made of purgative plants. They drank this tea to vomit and cleanse themselves because they believed that their contact with European society was corrupting their souls and spirits. In general, however, sacred plant use in this country has usually happened in secret, in very quiet and discreet ways, because it's not a practice that was commonly accepted by the larger community.

One notable exception to that has been the use of peyote within the Native American Church. Peyote is a cactus that grows in a very small part of Texas and in the northern part of Mexico and, as I mentioned, it has an indigenous tradition of use going back many centuries in northern Mexico. It came to this country and began to be used by Indians in the Southwest in the late 1800s.

In the early 1900s, the pan-tribal Native American Church, which gave a formal religious context to the use of this traditional plant medicine, was founded, and despite a number of legal threats over time as the church grew, its members were able to continue this religious practice, and laws were eventually passed that officially allowed it. When the current version of the Controlled Substances Act, for example, was written, an exemption for the use of peyote within the religious ceremonies of the Native American Church was included. This established a legal precedent for the legitimate religious use of an otherwise controlled substance.

In around 1965 a group of people in New Mexico founded the Church of the Awakening, which employed many different practices, including meditation, fasting, and prayer. They also used peyote in some of their ceremonies, and they applied for recognition by the government in order to be able to legally practice their religious faith, but their application was denied. The government asserted at the time that the substance was untested and might not be safe. This was a bit odd since for decades the government had allowed Native Americans to use peyote. Logically, I suppose, they were in effect admitting they didn't care about the potential health risks to Indians (or didn't really believe the health risk was real).

The problem of what to do when individuals or groups demand to be able to obey their religious beliefs when those beliefs go against the law is something the courts have had to wrestle with throughout our history. Cases against some Mormons who claimed that polygamy was a part of their religion are a well-known example. In that case, the courts ruled that since one could be a Mormon without being a practicing polygamist (the modern Mormon Church now forbids it), polygamy was not essential to Mormonism and could be outlawed without impinging on religious freedom. So the issue of how central to a religion a practice was became one important factor.

Eventually, over time, a body of law about when the government could legitimately interfere with the religious practice of its citizens developed. That standard was defined by a two-fold test: the first element was that the government had to demonstrate that it had *a compelling interest* in interfering with a person's religious practice. In the case of polygamy, for example, the government argued there was an interest in preserving the nuclear family.

The nuclear family, it was argued, was the foundation of an organized society, and if polygamy were allowed, it would lead to an undermining of the social fabric. An undermining of the social fabric could not be allowed: this was the government's compelling interest. The second element of this test was that in this attempt to balance the government interest in maintaining social order, with personal freedom, when the government became involved in curtailing people's religious freedom, it had to do so with the least restrictive means possible.

In 1990, a case related to the use of peyote within the Native American Church came before the U.S. Supreme Court. Although this case didn't directly relate to the use of the sacrament, the court used it as an opportunity to completely redefine the law as it related to religious freedom.

An Indian by the name of Al Smith worked in a drug treatment center and was a member of the Native American Church. Smith was told by his boss—a practitioner philosophically aligned with the twelve-step program who believed that the use of any substance was contrary to sobriety—that if he, Smith, wanted to continue his counseling work at the drug treatment center he would have to quit the Native American Church. Smith refused and was fired. He sued, and the case ultimately made its way to the Supreme Court.

There's a book called *To An Unknown God*, which documents the story of this case (Al Smith versus the Oregon Employment Division) and its magnitude and significance. The Supreme Court's ruling in this case changed the previous standards about the authority of the state in religious liberty cases. The Supreme Court ruled that if a law is generally applicable (i.e., it applies to everybody), and it's neutral in that it's not targeting religion specifically, then that law is valid. If that law has the unintended consequence of harming people's religious faith, so be it.

The Court essentially said that the previous degree of religious freedom was a luxury that our democracy could no longer afford. Justice Scalia wrote the opinion and was joined by four other justices. I believe that it was a five to four decision. The late justice Harry Blackmun wrote the minority dissent, which read:

This Court's decision effectuates a wholesale overturning of settled law concerning the religion clauses of our Constitution. One hopes that the Court is aware of the consequences and that its result is not the product of overreaction to the serious problems the country's drug crisis has generated. This distorted view of our precedents leads the majority to conclude that the strict scrutiny of a state law burdening the free exercise of religion is a luxury that a well-ordered society cannot afford, and that the repression of minority religions is an unavoidable consequence of democratic government. I do not believe the founders thought their dearly bought freedom from religious persecution a luxury, but an essential element of liberty and could not have thought religious intolerance unavoidable, for they drafted the religion clauses precisely to avoid that intolerance.

But the Court's opinion was written in such a way that it left open the possibility that Congress could legislate more stringent religious protections even if the Court wasn't willing to concede these protections necessarily existed within the Constitution. And Congress did in fact pass a piece of legislation in 1994 called the Religious Freedom Restoration Act. This Act was passed by unanimous voice vote in the House of Representatives and by ninety-seven to three in the Senate. And it is thanks to this law that we appear to have gained a victory in our legal case against the government of the United States.

Prior to this law being passed, there was a case in New Mexico in which a man named Bob Boyle, a non-Indian member of the Native American Church, was arrested for sending a shipment of peyote from Mexico to himself in New Mexico. The presiding judge, a perceptive and courageous man, now deceased, wrote a scathing opinion that rebuffed the government's attempt to prosecute this man. He wrote:

The government's war on drugs has become a wildfire that threatens to consume those fundamental rights of the individual deliberately enshrined in our Constitution. Ironically, as we celebrate the two-hundredth anniversary of the Bill of Rights, the tattered Fourth Amendment right to be free from unreasonable searches and seizures and the now frail Fifth Amendment right against self-incrimination or deprivation of liberty without due process have fallen as casualties in this war on drugs. It was naïve of this court to hope that this erosion of the constitutional protections would stop at the Fourth and Fifth Amendments, but today the war targets one of the most deeply held fundamental rights—the First Amendment right to freely exercise one's religion.

One of the consequences of the privatization of prisons in the last two decades is that there is now a large and influential private business whose commodity is incarcerated human life. Like any business, it wants to be a growth industry and, with violent crime on the decline, non-violent, non-dangerous drug users are a key to the steady flow of bodies to incarcerate. So the pressure to keep locking up more and more people in the name of protecting our society from dangerous drugs is very high. And while there certainly are some substances that are dangerous, clearly the way to heal society from their use is not through the massive incarceration of non-violent users. For substances that are not dangerous at all, or, like the ones we use in our religious practice, that are actually of potential benefit to individual health and consciousness, it is particularly insane.

What we have taken a stand for is the right to be able to receive nature's gifts from her and to use them in a disciplined, safe, structured, reverential, and sacred way. For more than forty years this has been the case in modern Brazil, and it has also been the case in the Amazon region through the centuries. The people I met in Brazil who had been drinking this hoasca tea for twenty, thirty, forty years, were lucid, open-hearted, humble, and gentle. They had wisdom to share and healthy families with children whose eyes sparkled with happiness and peace. Our government has taken a position of prohibiting the use of a substance that actually seems to provide great social benefits to its ceremonial users.

Another significant case involving religious freedom and plants involved an individual who wanted to create a church built around the use of marijuana. The case, which was decided in the 10th Circuit, Wyoming District Court, wrestled with how to define what a religion really is and what criteria to use to determine when someone is legitimately involved in a religious practice (as opposed to trying to claim a religious exemption when in reality someone is just trying to use religion as a cover).

The court decided that a religion has to have teachings that embody ultimate ideas about life, an understanding of cosmology, a way of relating to the world, metaphysical beliefs, a sense of transcendence, a sense that there is something beyond the physical or the mundane world, and some element of a higher power (not necessarily a God: they were very careful to not define specifically what people had to believe). There needed to be a moral or an ethical system, a code of conduct of how to live in the world in terms of practicing virtues. The judges said these are elements that are common to all recognized religious systems. It wasn't merely enough to say, for example: "We worship the marijuana plant and our practice is to smoke it all day long," which seemed to be the case in this particular instance. The judges ruled that there needed to be a more comprehensive body of belief in order for it to meet the legal standard of what constituted religion.

The court also said that one invariably finds certain accoutrements within authentic religious communities, and they listed them. These included: a founder or leader or prophet or teacher, someone who brought a message of wisdom or knowledge of life to humanity to begin the tradition; recorded writings that constitute the core principles or teachings of the faith; gathering places and set times where typically the community of believers meets to conduct rituals; keepers of knowledge, people who are trained to transmit the teachings; specific ceremonies and rituals; a body of laws and behaviors; days of special significance that are unique to that tradition; and often specific dietary rules.

Every one of these elements is present in the UDV. It has a master, a teacher who brought forth this way of gaining the knowledge of the divine. He was a rubber-tapper who encountered this practice of using hoasca as a sacrament in the rainforest in the north of Brazil. There are laws and writings that are read at the beginning of each one of our rituals. There are temples that have been constructed in over one hundred places all over Brazil where communities of one hundred and fifty to three hundred people gather regularly to participate in ritual.

There's a trained priesthood—a group of *mestres* who have the knowledge of the spiritual tradition and who transmit those teachings. There is a very defined structure in terms of how the sessions are conducted and how the teachings are transmitted. There are holidays at which we gather for the purposes of remembering events that happened in the establishment of our religion. We certainly meet all the criteria described by this ruling of the Wyoming District Court.

In Portuguese, União do Vegetal literally means "the union of the plants." There are two Amazonian plants that are brewed together to create the sacred tea we use in our religious ceremonies. One is a vine (*Banisteriopsis caapi*) and the other is a leaf from a small bush (*Psychotria viridis*). On one level the name União do Vegetal describes the union of these two plants to make our sacrament, but it also describes other "unions": achieving a state of union with the sacred through the plants, the union of community, the union of our human consciousness with this gift from the natural world, and the union of human beings with nature.

These are all elements in the União do Vegetal's teachings. For us, the full realization of our potential begins by acknowledging that nature is sacred, and as we humans recognize that and become instruments for the expression of divine nature, we realize that we too are part of nature and sacred.

The story of the history of the origin of our sacred tea, according to the teachings of the UDV, has been a guarded secret for generations. On certain occasions it is told within our rituals. It is said that it goes back thousands of years, before the early beginnings of the Inca Empire, and that the knowledge of these two plants, how to combine them and how to use them in ritual, spread throughout the Amazon forest. Today there are still many tribes that retain this knowledge—or elements of it. In the late 1930s and 1940s, as the war generated a large demand for rubber, thousands of men from all over Brazil came to work in rubber plantations in the Amazon, and a number of them learned the use of the tea from the Indians.

One of these was José Gabriel da Costa, who we speak of as Mr. Gabriel. He is the founder of our tradition, which is now in one hundred cities, villages, and regions all over Brazil, and has now spread to the United States as well as a few countries in Europe.

These rubber-tappers lived under conditions of near slavery, getting up at two o'clock or three o'clock in the morning and working all day for almost no pay. They would walk through the forest gathering the sap that dripped from the rubber trees (before rubber was synthesized, this was the only way it was manufactured). The world military industrial complex at the time was dependent upon this terrible slavery, but at least some good indirectly came of it: Mestre Gabriel encountered the use of this sacred tea, and he began gathering disciples and revealing the mysteries of nature to them through the tea, and he founded a temple. Today there are União do Vegetal temples in more than one hundred places around Brazil.

In UDV ceremonies several hundred people in a community come together to prepare the tea, working day and night in a climate of love, peace, and harmony. Our understanding is that everything that goes on in the preparation of the tea is recorded within the tea. Since we use the tea as a tool for spiritual illumination and bringing peace, wisdom, and understanding into our consciousness, it has to be prepared in an atmosphere that reflects that. The plants are boiled in the essence of life—water—into which they release their mysteries, knowledge, and wisdom.

The UDV came to the United States in 1987 after an American physician and his companion (who were traveling in the Amazon to help bring medicine to forest peoples) encountered the UDV. They were deeply impressed and asked the UDV to send two teachers to the United States, and the church hierarchy agreed. In 1990 I made my first trip to Brazil and I worked with a few other people to start to bring more UDV *mestres* from Brazil to begin to hold ceremonies in this country.

We officially incorporated as a church in New Mexico in May 1993, after the Religious Freedom Restoration Act became law. For six years we had regular sessions and meetings, and the UDV expanded into a number of different cities in this country. In May of 1999 agents of the U.S. Customs Service and the FBI came to my office doing what they called a "controlled drop." They delivered a shipment of our sacrament that was sent to us from Brazil and, after I accepted and signed for the delivery, a SWAT team of twenty to thirty armed agents with dogs came in.

They stayed for about eight hours, took all of my computers, personal records and forty-thousand documents from my office, and began an investigation of the UDV in this country. They sent agents out to five states to see if they could gather information they could use against us in a criminal case. They convened a grand jury. We ultimately filed our own lawsuit against them after eighteen months of trying to negotiate an understanding with the representatives from the Department of Justice. They showed absolutely no interest at all in working out some kind of agreement with us, so the only option we had was to sue them.

Our lawsuit was very carefully prepared. It was something I had long been thinking we might have to deal with. The Department of Justice put together a team of forty lawyers to handle our complaint, from their criminal, civil, constitutional, international law, and public health divisions, the Food and Drug Administration and the Drug Enforcement Agency (DEA), and all this for a church with maybe some 120 peace-loving members in the whole country! We had two lawyers—Nancy Hollander and John Boyd—part of a small law firm in Albuquerque. Fortunately for us they're deeply committed and really good lawyers.

We filed our complaint (with the Department of Justice), based on constitutional law and the principle of religious freedom, on rigorous studies that proved the medical safety of our sacrament, and especially on the Religious Freedom Restoration Act. We accused the government of violating our First Amendment, Fourth Amendment, and Fifth Amendment rights, with unreasonable search and seizure, with denial of due process for confiscating our religious sacrament and refusing to return it despite repeated attempts, and with violation of the Equal Protection Clause of the Constitution.

We filed this last part of the complaint because there's another religion in this country (the Native American Church) that was granted the right to use an otherwise controlled substance within their religious ceremonies, which has not been interfered with by the government for decades. We felt *that* established a significant legal precedent that required the court to give very serious consideration to our religious use of our sacrament as well.

Our sacrament is considered to be a controlled substance because it contains small amounts of molecules of dimethyltryptamine, a substance that's actually produced in the human brain. It's part of our nature. The government's claim that this substance, naturally found within all humans, is somehow the equivalent of heroin or crack cocaine is absurd. It is our view that every element in our sacramental tea is, in fact, part of our nature. When you drink it you're not receiving something foreign into your body. Rather you're synergizing your own nature so that you see, hear, feel, and think more clearly.

Besides U.S. law there's also a body of international law that affirms the right of people to practice their religion freely without interference from the state. As a result, we brought several U.N. declarations into our case (which is why some of the forty lawyers involved were from the international division—they had to deal with the fact that we were claiming the U.S. Government had violated these treaties). It also got very technical, as the law usually does. We brought up the improper application of the Administrative Procedures Act in their confiscation of the tea. We requested a court order from a federal judge stating that the government was no longer permitted to interfere with our religious practice and that they had to return the tea and order the Customs Service and the DEA to allow our shipments of tea to enter the country.

The government's response was that "surely neither the Controlled Substances Act nor the Religious Freedom Restoration Act required the government to wait until it had 'a full-blown drug epidemic' on its hands before it attempted to stem the tide of usage." They argued it was the government's responsibility to protect the public health, and insufficient studies had been done—there might be dangers associated with this tea. The fact that it's been used for centuries within the Amazon and for decades within modern Brazilian society, and that there were organized communities of urban, sophisticated people who had been consuming it for forty years with no adverse effects and that there had, in fact, been good studies showing its harmlessness didn't seem to matter to them.

It's instructive to compare their attitude to that of the Brazilian government. When the Brazilian authorities became aware of the growing use of hoasca in their country in religious rituals, they sent a commission of doctors, psychologists, theologians, and public policy people to interview the people using the tea to find out what the effect was on their lives. Some of the commission members even tried the tea. As a result of their report, hoasca was formally legalized for use in religious rituals in Brazil in 1992.

The U.S. government lawyers desperately searched for and came up with as many arguments as they could muster. Most were obvious, and we were prepared to counter them, but some caught us off guard. One of the most initially problematic arguments for us was that the United States had a responsibility to honor its treaty responsibilities, and apparently the United States was a signatory to a 1971 treaty called the International Convention on Psychotropic Substances that (at least the Government claimed) made illegal the use of dimethyltryptamine in any form.

This argument is deeply ironic, in that this case deals with knowledge originating with a millennial history among indigenous people, and the U.S.'s record of honoring treaties is far from stellar. That's especially true for treaties with Indian nations, which were routinely flouted and violated throughout our history.

But this was a difficult wrinkle in the case. When I went to the website of the National Narcotics Control Board and had my first look at this treaty, it appeared, on its face, that we had come up against a barrier that was going to be very difficult to get through. We contacted an expert in treaty law, who explained to us that all treaties have a record of commentary attached to them that is the equivalent of the congressional record of how a law is formed. Often the law says one thing at first glance, but if you study the congressional debate that lead to the passage of the law, you might see its intention in a different light. He explained that in the commentary to this treaty there might be grounds for legal protection in our case.

We looked for a complete copy of this treaty, one that would contain the commentary which might prove useful to us. We searched far and wide—in New York, in Washington, D.C. Nothing. We ended up having to send somebody to Austria, where there's a library of international agreements, to finally get a complete copy of this convention. I spent hours looking through it to find something. One night, going through line after line after line, I came across this paragraph:

> Schedule One of the treaty does not list any of the natural hallucinogenic material in question but only chemical substances which constitute the active principles contained in them. Neither the crown, fruit, button of the peyote cactus . . . nor psilocybe mushrooms themselves are included in Schedule One, but only their respective active principles—mescaline, DMT, and psilocybin.

Later I found another section that clearly stated that the treaty did not intend to target historical use of plant materials within ceremonies of magical and religious rights by clearly defined groups. Even with the hysteria of that era surrounding the spread of psychoactive substances (as this international law was drawn up in 1971), the signatories recognized legitimate use within magical and religious rights by clearly defined groups of people where there had been a history of that use. So much for the government's supposed treaty obligation.

The government's case at its core, in the end, was based on three compelling interests: the treaty I just discussed, public health, and the risk of diversion of our sacrament outside of its religious context. Their arguments about health risks were weak. We offered in evidence a 1992 study by an international consortium of scientists, botanists, biologists, chemists, and psychiatrists who came to the UDV to study our sacrament, and published their results in several medical journals. They found no health problems associated with hoasca use among UDV members. And this was the only U.S.-based study done to this point, so there weren't any competing studies that showed any harm the government could invoke.

The second issue was the risk of diversion of our sacrament outside of its religious context. We had a former Justice Department official from the Criminal Division's Narcotic and Dangerous Drug Section testify on our behalf. When asked to evaluate the risk of our sacrament being stolen out of its religious context and abused recreationally outside of the UDV, he said, "people who are hungry are not going to break into a Catholic church and steal communion wafer for food. If they want to steal bread, there are far larger lots of bread to be found in other places."

The foundation of our argument, in terms of the risk of diversion, was that there were so many other ways that people who are looking to get high could get high, the likelihood of thieves seeking out relatively small amounts of an exotic tea—which is very unpleasant tasting and hard to prepare and make use of properly without expert guidance—is exceedingly low. The truth is that if you use this sacrament outside of the context it's meant to be used in, it's not going to be fun.

Finally, on August 12, 2002, Judge Parker, the Chief Justice of the Federal District of New Mexico, issued his ruling. It said that the government had not shown that applying the Controlled Substance Act's prohibition on DMT to the UDV's use of hoasca furthered a compelling interest, that the government had not proven that hoasca posed a serious health risk to UDV members who drink the tea in a ceremonial setting, and had not proven the risk of any significant diversion of the substance to non-religious use.

AN UPDATE ON THE UDV CASE

Jeffrey Bronfman's presentation above took place in October 2004. As developments occurred in the case, we asked him for updates. The first update came in November 2005; the second on February 21, 2006, when the U.S. Supreme Court issued its ruling:

On November 1, 2005, the United States Supreme Court heard Oral Arguments on the U.S. government's appeal of the preliminary injunction that had been granted to the UDV in December of 2002. The injunction prohibited the U.S. government and its agents from interfering with the UDV's importation, distribution, and ceremonial use of its religious sacrament.

The government's appeal of this order was its third, having already unsuccessfully appealed the Federal District Court's decision to the 10th Circuit Court of Appeals twice before. Some Supreme Court analysts and legal scholars considered this case the most important religious freedom case the Court has accepted in decades; others since the Continental Congresses that led to creation adoption of the U.S. Constitution more than two hundred years ago.

In support of the UDV's petition for religious liberty and tolerance of its central, but not well understood religious practice, dozens of religious and civil liberties organizations independently authored or signed legal briefs submitted to the U.S. Supreme Court. These organization included The Catholic Bishops of North America, The Joint Baptist Committee, The National Council of Evangelical Christians, The Presbyterian Church of The United States, The American Civil Liberties Union, and The American Jewish Congress.

THE FINAL OUTCOME

On the morning of February 21, 2006, the Supreme Court of the United States published a unanimous ruling in favor of the UDV and its religious use of its sacrament hoasca. A website containing a detailed analysis of the legal action brought by the church, containing all of the different court decisions, was established on the Internet at www.udvusa.com.

Editor's Note:

The outcome of this extraordinary, historic case, widely viewed as the most significant Supreme Court ruling to date on the thirteen-year-old Religious Freedom Restoration Act (which places a very high burden on the government in seeking to restrict religious observance) does not, of course, sanction shamanic plant use outside of the specific confines of UDV religious practice. However, it is clearly of enormous significance for anyone with a deep respect for the potential "entheogenic" properties of certain plants with a long history of sacred use.

This presentation took place at the Bioneers Conference in 2004, with an update in 2006.

EPILOGUE

The Madness of the War on Drugs: A Tragically Flawed Policy's Ecological & Social Harms

Michael Stewartt and
Ethan Nadelmann, Ph.D.

Michael Stewartt: Twenty years ago I was living in Grand Junction, Colorado, and I bought the only house I've ever purchased in my life. It was a tiny little house that I really loved, but the exciting thing about it for me was that it sat on half an acre. I had always dreamed of being a farmer. It was in the desert and my land was bare, but the property had access to ditch water, so I could use plenty of irrigation water. I was really excited. I went to work on a quarter of an acre to turn it into a huge garden. My next-door neighbor had one of the most beautiful gardens I've ever seen.

So I went to work. I rented a Rototiller. The ground was really hard and I had to move rocks around. One day I was digging irrigation ditches by hand, sweating buckets and working hard, and that neighbor, Frank, who was seventy-two years old at the time, called me over to the fence that we shared and said, "Hey, you're obviously trying to get a garden going here."

When I replied that that was so he said, "Well, I think it may be helpful for you to know that you may have a problem because the fellow you bought this property from didn't want to mess with the weeds on this part of the lot, so he sprayed it with a combination of weed killer and kerosene. The weed killer is long gone; it biodegrades quickly, but there's probably still kerosene in your soil."

He gave me advice: "Look, as soon as the ditch water comes, here's what I think you should do: run it onto your plot and just let the water stand there as much as you can. Let the water try to push the kerosene that's still in the soil down so the plants can't get to it." So I did just that. For almost two weeks I had what looked like a future rice paddy in the desert. I had it under water.

Time passed, and I let it dry just enough, got my rows together, and planted onions, squash, beans, carrots, tomatoes, and corn. I was very excited about it. Six to eight days later, all these little sprouts came up looking very healthy! I thought it was going to work. About three weeks later, every single plant turned brown and died. You can imagine how devastated I was. The plants had grown just enough for the roots to get down to where the kerosene was still in the soil.

Frank had told me that the guy I'd bought the property from might have used a gallon of kerosene on this quarter of an acre. The Ministry of the Environment in Colombia estimates that people producing illicit cocaine dump over six million gallons of kerosene onto the ground and into the waterways of that nation every single year. In my case, my story had a happy ending: with Frank's good tutelage and support and, after a tremendous amount of work, a year later I had a wonderful garden. The Earth can heal, but one of the biggest problems no one talks about in the drug war is the massive toxicity caused by the chemicals that both the narco-traffickers and the authorities use.

A recent United Nations report estimates that in Peru, Colombia, Bolivia, and Brazil, every single year over seven million gallons of combined acetone, sulfuric acid, ethyl-ether, and ammonia are dumped onto the ground and into the water courses. We do a lot of work on forest conservation, and one of the main tools we use to educate key people is to fly them over the land. You can see the extent of devastation much more clearly from the air. In those aforementioned countries the hillsides in drug-growing areas are ravaged to an unbelievable extent. I know because I've flown over most of them.

The National Academy of Science, using data they've received from NASA, estimates that illegal drug crops are the third leading cause of global deforestation. Millions and millions of acres of remote areas are being cleared of forests and ultimately made toxic so that people can grow crops that we, in our culture, primarily in the United States, have decided are illegal: marijuana, opium poppies, coca. We need to understand that it is because of our decision that these plants have to be illegal and the growing of them therefore repressed that we get this astonishing deforestation and toxification in remote wildlands.

Prohibition of a highly desired commodity always creates a black market that results in high prices, which, in turn makes the incentive to grow these plants irresistible. The repression of the growers leads them into more pristine, remote regions, where the drug gangs often subjugate, expropriate, and oppress the local, indigenous populations. No one would normally go to high mountain country where the soils are poorer, where it's tough to get irrigation water for the crops, tough to get the crops to market, and far from roads and transportation, but the illegality and resulting inflated value of the crop change the equation.

Here at home we feel the impact of the drug war in a different way. We live in one of the most violent societies on the planet, and much of the violence is the result, not of drug use per se, but of the insane profits that can be made in the drug trade, because those drugs are illegal. Just like in the days of prohibition, the illegality of the substances causes more harm than the substance itself.

A close friend of mine in Albuquerque was recently awakened by gunfire. A drug gang had come by, intending to shoot at their rivals in a turf battle being fought over which gang had the right to deal drugs in the neighborhood, but they had fired into the wrong duplex. They hit a twenty-eight-year-old woman who was holding her two-year-old daughter in her arms. The first round that came through her wall went right through her thigh. The second round went right through her baby's head. Her daughter died instantly in her arms. That is not such an unusual story in our country. People who have nothing to do with the drug trade are gunned down over and over in this country because of the illegality of these drugs.

We at Wild Angels did a lot of work in land conservation in Mexico and had some success in convincing the government to protect some key forests. We've also been blessed with some major successes in contributing to saving millions of acres of threatened forests in Central and South America and in this country, but the drug war is eroding all the work we and many other groups have done. Some of the most isolated groups with unique cultures are most at risk. In Mexico, the renowned and much-studied Tarahumara Indians have seen their lives brutally disrupted by this drug trade; they are often forced into near slavery by armed gangs with ties to the local authorities. And they are far from alone. The same pattern is occurring to indigenous peoples across Latin America.

Wherever people in the Americas work the soil they are suffering the consequences of the U.S. drug war. Ask them what can be done, and they all have the same answer: "You have to legalize it." That's the only way to stop all these horrors. It's the same pattern the country experienced during Prohibition. When potential profits are obscenely high, people will do astonishingly horrible things, and the fabric of society is corrupted by illicit money.

In this country, we may not love the chemical industry. We may not always get from the chemical industry what we'd like in terms of their citizenship on this planet, but we are able, by and large, to exert some control over them. They can't just dump huge quantities of toxins anywhere at their whim because they're legal entities with legally monitored products working within a framework of the law. That's what we gain through legalization and control. But absolutely nothing stops drug cartels from poisoning the land and water with toxic chemicals because they operate completely outside the law.

Right now the United States spends over $75 billion a year on the drug war, including the cost of incarcerating people for drug offenses. I think we can all come up with far more productive uses for that money: education, intelligent foreign aid, alternative fuel development, and protecting the environment, for instance. The government's approach to drugs is absurd: a recent ad linked teenage drug use with terrorism. Partnership For a Drug Free America runs ads on television all the time that link the use of marijuana with rape, murder, and so on.

Teenagers are not morons. They notice that if they smoke a joint, they don't get violent or become a risk to themselves and others, unlike with alcohol inebriation, so when they're told marijuana leads to all these horrible things, and their experience demonstrates that it's not true, what is left for us to do to convince them that heroin and methamphetamine really are bad for them? If we had educational programs about drugs that presented the complex, nuanced truth, we'd all be much better off and wind up with lower drug use.

Nearly half of all the people we incarcerate in this country are in jail on drug-related charges. Most of them have no connection to violent crimes. They haven't caused any harm except, perhaps, to themselves. The United States is now the world's biggest jailer, with the highest per capita rate of incarceration of any nation, even ahead of Russia and China. Between today and 2010, we will be releasing from our prison systems over three million people who will have done sentences of at least five years. A lot of those people will have done some hard time—time in solitary confinement, in a jail cell by themselves with no contact with other people twenty-three hours a day. This is a near foolproof recipe for creating violent crime.

I've been in jail five times for nonviolent civil disobedience. The longest stint I ever did was one week, which is, by prison standards, nothing. But even a week of being in a tiny cell with super bright lights on twenty-four hours a day, eating horrible food, and having a camera in the upper corner of the ceiling watching me even using the toilet, made me angry. I came out more pissed off than when I went in.

One goal of incarceration used to be rehabilitation; today it's just punishment. We put people in terrible conditions, and when they come out they tend to be deeply angry and prone to violent acts. In the next eight years, we're going to release over three million such people who've been in jail for at least five years. We can expect to pay a high price in violence in return. President Jimmy Carter once said: "Penalties for the possession of a drug should not be more damaging to the individual than the use of that drug itself." Our prison system today is very good at taking relatively harmless drug users and creating sociopaths.

The madness of the drug war, already causing such social and environmental harm, will soon be devastating whole ecosystems. The U.S. government-backed effort to eradicate drug crops involves aerial spraying with potent herbicides. But they're looking for a magic bullet. They want something that will just eradicate everything. Mostly they've used a chemical called glyphosate in Latin America under Plan Columbia. It's a very harsh herbicide, but the government wants to use a much stronger, much less biodegradable herbicide produced by DOW Chemical that is poisonous for years. And they're looking at things like genetically altered bacteria and spreading parasitic fungi. These are profoundly stupid and frightening ideas. Time is short for us to turn things around for this Earth, and we in the environmental movement must understand that the war on drugs is one of the most ecologically and socially devastating policies of our government, and we must join the struggle to end it.

Ethan Nadelmann: I'm envious of the environmental movement. It's really big. It loses some battles, but it's still big and influential; it's had a lot of success in raising our awareness about how human beings should relate to the environment around them. Those of us concerned about the way this country and other countries are dealing with drugs are trying to build something similar: a new consciousness and a new movement that will change the way people think about drugs, drug policy, drug users, drug addicts, and drug markets, and the ways that governments deal with all this.

Michael mentioned that the United States has the highest per capita incarceration rate in the world. If it were another country, we would immediately think of this statistic as evidence of insensitivity to basic human rights. But when we look at ourselves, we don't tend to see it that way. I've often heard that the United States has 5 percent of the world's population yet consumes 25 to 30 percent of the world's energy resources. Well, we also incarcerate 25 percent of the world's prison population. There are close to nine million people behind bars around the world, and roughly 2.2 million of them are in the United States.

The frustration for me is how hard it is to get people to care about this. We've gone from a half a million people behind bars in 1980 to quadruple that number in just twenty-five years. We've gone from fifty thousand locked up on drug charges back then to almost half a million people locked up solely on drug charges today. We lock up more people in America for breaking a drug law than Western Europe locks up for everything, even though their total population is roughly 30 percent more than ours. The U.S. prison industrial complex has grown enormously. And those incarcerated on drug charges are preponderantly nonviolent. Many of them are just people who want to put a drug in their bodies, and most of the small-timers who make and sell drugs do it the same way many of our grandmas and grandpas made bathtub gin or backyard booze sixty to eighty years ago.

Drugs certainly present a confusing picture. None of us who want to change the laws are suggesting we should allow drugs to be sold to children, or that the most potentially problematic substances be totally legal and unregulated. But we certainly want to decriminalize mere use and possession of drugs and stop giving nonviolent drug offenders harsher sentences than rapists and murderers. Many drugs are useful substances when used judiciously, and our attitudes toward them and laws regarding them have been very different during different historical periods. Heroin, for example, can be legally prescribed as a painkiller in the U.K.

and Canada. And a hundred years ago, heroin, opium, morphine, marijuana, and cocaine were all legal while alcohol was being criminalized in states and counties around the country in the years leading up to Prohibition. In a few states, women could not legally smoke cigarettes. We had a different moral sensibility about which drugs were "evil" and which were okay. Perhaps more people used those drugs per capita one hundred years ago than is the case today, but there was far less of a "drug problem" than grips society today.

In America we are bombarded with the ideal of a "drug-free society." In 1988, Congress passed a resolution declaring that America would be drug free by 1995. How silly can you get? There's never been a drug-free society in the history of human civilization except perhaps in the coldest climates where no mood-altering plants grow. How much respect can you have for political leadership that makes such ridiculous pronouncements?

But, one might ask: should we not aspire to, and aim for, a drug-free world? I would argue we should not, because when you create an unachievable, totally unrealistic objective, you set destructive forces in motion, whip people into a frenzy, and wind up in a perpetual war that cannot be won. We (i.e., our governments) spend enormous sums of money to incarcerate very large numbers of human beings, and go to absurd lengths to reduce the amount of drugs or the number of people using drugs, but never stop to notice that we're creating far more monstrously destructive problems than the ones we started with.

We need a different framework for dealing with drugs. We need to begin to understand that the challenge is not how to get rid of drugs or how to keep them at bay. The challenge is how to learn to live with drugs; how to learn to live with this wide range of plants and chemicals so that they cause the least possible harm and bring the greatest possible benefit.

The myth that we should be drug-free is based upon its own myth: the idea that somehow we all pop out of our mothers' wombs as perfectly balanced chemical creatures. But there is a wide range in people's brain chemistries. I've heard that some users of heroin and some patients who respond well to Prozac use the exact same words to describe their experience of those drugs: "This is the first time I've felt normal." Why do these drugs "work" for them but not for others? One reason is the differences in our chemical make-ups.

A few decades ago, it was discovered that we human beings produce our own natural opiates: endorphins. Neurologists still admit to having a very rudimentary understanding of brain chemistry and the physical basis of our thoughts and emotions. Perhaps the reason some people respond to heroin is because they're born with an inadequate capacity to produce natural endorphins, so heroin or other opiates make them somehow feel "normal."

So, while there's no question heroin can cause problems, it's chemically in the same family as those endorphins in our body. It's not some totally alien, "unnatural" substance. In fact, if you were to somehow take the endorphins out of your body and then put them back in, you'd likely be violating the law. And ten years ago it was discovered that we all have receptor sites for cannabinoids as well. In other words, we are all "potheads" whether we smoke it or not.

Perhaps a more realistic approach to drugs for most people is to seek a balance between the chemicals that lie within our brains and those that lie without. Keep in mind that the vast majority of people in most cultures (with the exception of some indigenous and other societies that have been devastated by rapid social change) find ways to achieve some kind of balance. Almost everyone uses one psychoactive drug or another—it just depends upon how narrowly or broadly one defines the term. Most people do no harm to themselves or anyone else as a result of their drug use, and most people who misuse drugs at some point in their lives eventually stop doing so.

Drugs can, of course, be dangerous, some more than others, some even deadly. Sensible education and regulation can play a constructive role in reducing drug misuse and its negative consequences. But it is a fundamental mistake to confuse criminal prohibition with sensible regulation. The horrific consequences of the war on drugs today stem, in good part, from the misguided notion that the criminal justice system must play an essential role in protecting us from certain drugs. That system has proven to be a crude tool, and one whose principal actors no longer know, much less respect, the original public health rationale for drug control.

There is a core principle, which lies at the heart of the drug policy reform movement. It is that none of us deserve to be punished or discriminated against solely for what we choose to put into our bodies. This principle applies equally to the social or recreational user and to the addict.

There is no legitimate basis in science, ethics, or even religion for the law to discriminate between the responsible consumer of alcohol and the responsible user of marijuana, cocaine, or heroin. Nor is there any legitimate basis for the law to discriminate between the alcoholic and the heroin or cocaine addict, absent harm to others. But what about people who are impaired on the roads or in the workplace, one might ask? They are not excused by this principle. Punish people for the harms they do to others, whether or not they are under the influence of drugs—and do not accept addiction as an excuse for harming others. The flip side of individual freedom, in any just society, must be individual responsibility.

The drug policy reform movement may be tiny compared to the environmental movement, but we are growing and increasingly winning political battles, especially at the state and local level. We succeeded in winning passage of California's medical marijuana initiative (Proposition 215) in 1996, and then drafted and won additional medical marijuana initiatives in Washington, Oregon, Alaska, Colorado, Nevada, and Maine.*

---

*California's Proposition 215, also known as the Compassionate Use Act, was designed to protect seriously and terminally ill patients from criminal penalties for using marijuana medically, but only people with their doctor's recommendation to use marijuana in medical treatment can take advantage of Proposition 215 as a legal defense against marijuana charges.

We're now trying to fight the DEA's attempts to ignore the wishes of the electorate. Recently DEA agents raided a medical marijuana hospice in Santa Cruz, California, where the vast majority of the patients are terminally ill. The agents who were armed with automatic weapons went so far as to handcuff a paraplegic woman—although I should also note that a few agents were heard wondering out loud what the hell they were doing there. My organization is now suing the federal government in this case.

The current administration is about as reactionary as possible on this issue; they are about as unenlightened a bunch on drug issues as one could find. But as our movement continues to grow and gain influence, things will change. We have an office in Washington that's beginning to target Congressional representatives and Senators from states where majorities have approved medical marijuana initiatives, asking them whether it's appropriate for the DEA to be violating their state laws, ignoring the voters' will, and persecuting medical marijuana patients.

We also drafted, campaigned for and won Proposition 36 in California in 2000 to prohibit the incarceration of people arrested for nonviolent drug possession offenses and double state funding for drug treatment. It passed with 61 percent of the vote in California, even though the governor, the attorney general, and nearly every police chief opposed it. It's probably the most significant sentencing-reform legislation since the repeal of Prohibition. It will result in tens of thousands fewer people being incarcerated in the nation's biggest state on drug charges, and saving taxpayers hundreds of millions of dollars.

The hemp issue is another example of the madness of our obsession with drugs. Forty countries grow industrial hemp legally, and it's a business worth hundreds of millions of dollars, but you can't grow it in America. Hemp has myriad uses as a fiber, a food source, and as a paper substitute—it's actually more durable than standard paper. George Washington grew hemp. It was the main source of rope and linens (and often paper) throughout much of the world until about 1850.

During World War II the federal Department of Agriculture put out a movie called *Hemp For Victory*. They wanted farmers to start growing it again—just a few years after the federal government had basically banned it—because the nation's supply lines of many other fibers had been disrupted. Hemp is basically marijuana, but it generally contains only tiny amounts of THC, and certainly not enough to make you high—even if you smoked pounds of it. U.S. Customs even tried banning birdseed imports from Canada because there may be some industrial hemp seeds in the mix. Don't these people have anything better to do?

The drug war in its current form is really a form of quasi-religious fanaticism. And if we study its history, we will find it has very racist origins. The first anti-drug laws in America were anti-opium laws in the 1870s and 1880s, directed at Chinese migrants to California and Nevada who, it was feared, would lure white women into their opium dens where they would be addicted and seduced. The first anti-cocaine laws were in the South, in the first decade of the last century, and were directed at African Americans. And the first anti-marijuana laws in America were in the 1920s in the Southwest and the Midwest and were directed at Mexican Americans and Mexican migrants.

Addiction is a behavior; it's not just about drugs. A gambling addiction is almost the same as a crack addiction. The dynamic of the phenomenon is almost exactly the same. There are all sorts of destructive addictions. Is crack devastating and destructive? Yes, it is, but the fact that it's illegal and people design their whole lives around getting it makes it far *more* destructive. I'm not advocating that all drugs be legalized and sold like alcohol or cigarettes, but there's no question that ending the current mindless, repressive, and inflexible prohibition policy would dramatically reduce the violence and corruption.

There are no easy answers. Millions of people struggle every day with decisions such as: "Do I or do I not put my kid on Ritalin so he can focus better in school?" "Do I or do I not take Prozac, or pain medications?" Let's refrain from simplistic, moralistic judgments and look for fundamentally humanitarian answers. I urge everyone who understands just how destructive and wasteful this war has been to make ending it a priority in your own life. We are building a political movement that follows in the footsteps, and stands on the shoulders, of the movements for civil rights, women's rights, gay rights, and respect for our environment.

This discussion took place at the Bioneers Conference in 2002.

Contributors

Marcellus Bear Heart Williams, a full-blooded Muskogee Creek Indian and a trained medicine man, is the author of *The Wind Is My Mother: The Life and Teachings of a Native American Shaman.*

Jeffrey Bronfman, a teacher, social activist/philanthropist, and president and managing director of the Aurora Foundation, is the legal and spiritual representative of the União do Vegetal (UDV) religion in the United States, and one of the plaintiffs in the UDV's historic lawsuit against the Federal Government. The outcome of this lawsuit was ultimately decided upon by the U.S. Supreme Court in February 2006. Jeffrey was also director of the Threshold Foundation and the Live Oak Fund for Change, was on the jury for the Right Livelihood Award, and was chairperson for the Congress of Spiritual and Religious Leaders at the 1992 United Nations Earth Summit on the environment held in Rio de Janeiro, Brazil. A student of the sacred for three decades, he has been initiated in the ceremonial traditions of many tribal cultures.

Katsi Cook, a Wolf Clan Mohawk, mother of six, grandmother of seven, is a maternal and child health consultant whose mission is to restore traditional birth and reproductive health knowledge to Native American families through the development of community-led and culture-based research, outreach, education, and clinical practice models. The founding aboriginal midwife of the *Tsi Non:we Ionnakeratstha* (the place where they will be born) *Ona:grahsta'* (a birthing place) Six Nations Maternal and Child Centre in Hagersville, Ontario, Katsi's work integrates environmental health research, aboriginal midwifery, and the democratization of scientific knowledge and systems of indigenous knowledge. A key figure in a campaign to evaluate and monitor the effects of PCB and other organo-chlorines on breast-feeding Mohawk women in the St. Lawrence River and Great Lakes Basin ecosystem, Katsi was also a professional member of the Interim Regulatory Council of the College of Midwives of Ontario, and is writing curriculum to assist in the training of a new generation of aboriginal midwives, doulas, and other birth-workers.

Wade Davis, Ph.D., one of the world's most renowned anthropologists, studied ethnobotany at Harvard and is the author of many books, including *The Serpent and the Rainbow, Penan: Voice for the Borneo Rain Forest, The Clouded Leopard, One River,* and *The Light at the Edge of the World.* While living among fifteen indigenous groups in eight South American countries, Wade created six thousand or so botanical collections. He continues to be a research associate of the Institute of Economic Botany of the New York Botanical Garden, and a Fellow of the Linnean Society, the Explorer's Club, and the Royal Geographical Society. He is an activist affiliated with a wide range of groups and NGOs including the Endangered People's Project, of which he is the executive director. He is a board member of the following organizations: The David Suzuki Foundation, Ecotrust, Future Generations, Cultural Survival, and Rivers Canada, and is currently Explorer in Residence at the National Geographic Society.

Alex Grey, an artist based in Brooklyn, New York, has achieved worldwide renown, especially for his extraordinary X-ray-like portraits of the human body's physiological and energetic systems, and for his search for a common mystical experience underlying all of the world's spiritual traditions. He has courageously and unhesitatingly acknowledged his deep debt to vision-inducing substances in helping to shape his artistic vision. His art has been exhibited throughout the world and is chronicled in three books: *Sacred Mirrors: The Visionary Art of Alex Grey, Transfigurations,* and *The Mission of Art.* He also co-edited a book about the conjunction of Buddhism and psychedelics, *Zig Zag Zen.* His work is on display at a unique gallery/community space, the Chapel of Sacred Mirrors, in the Chelsea neighborhood of Manhattan.

Charles S. Grob, M.D., a professor of psychiatry and pediatrics at the UCLA School of Medicine and director of the Division of Child and Adolescent Psychiatry at Harbor-UCLA Medical Center, conducted the first government-approved study of MDMA ("Ecstasy"). He was the principal investigator of an international biomedical psychiatric research project on ayahuasca use in the Brazilian Amazon. He is currently conducting an investigation of the effects of psilocybin on anxiety in terminal cancer patients. Dr. Grob has published numerous articles in medical and psychiatric journals and collected volumes, and is the editor of *Hallucinogens: A Reader* and co-editor (with Roger Walsh) of *Higher Wisdom: Eminent Elders Explore the Continuing Impact of Psychedelics,* published in 2005 by SUNY Press. He is also a founding board member of the Heffter Research Institute.

Kathleen (Kat) Harrison is an ethnobotanist, artist, illustrator, and photographer who researches the relationship between plants and people, with a particular focus on art, myth, ritual, and spirituality. She teaches at the California School of Herbal Studies, Sonoma State University, and the University of Minnesota (with field courses in Hawaii), and has done fieldwork in Latin America for thirty years. She is the cofounder and director of Botanical Dimensions, a nonprofit foundation whose aim is to preserve plant knowledge as it pertains to medicinal and shamanic usage. Kat Harrison's is a unique voice in ethnobotany. She brings deep integrity and fearlessness and a lovingly intense, profound intelligence to the study of medicinal and sacred plants and human cultures.

Francis Huxley, a widely traveled anthropologist who trained at Oxford University, is part of the illustrious intellectual tradition of the Huxley clan. He is the author of, among other books, *Affable Savages: An Anthropologist among the Urubu Indians of Brazil; The Invisibles: Voodoo Gods in Haiti; The Way of the Sacred; The Dragon: Nature of Spirit, Spirit of Nature; The Eye: The Seer and the Seen;* and with Jeremy Narby he is the co-editor of *Shamans through Time: 500 Years on the Path to Knowledge.* Francis has worked for half a century in highly diverse settings—among Amazonian Indians, Haitian voodoo priests, and with R. D. Laing and other radical psychotherapists and healers. He is an extraordinarily colorful character, one of the most offbeat anthropologists one could imagine.

Luis Eduardo Luna, Ph.D., was born in Florencia, in the Colombian Amazon region, in 1947. He studied philosophy and literature in Madrid and taught at Oslo University. He received his Ph.D. from the Institute of Comparative Religion of Stockholm University and an honorary degree from St. Lawrence University in Canton, N.Y. A Guggenheim Fellow and Fellow of the Linnean Society of London, he is the author of *Vegetalismo: Shamanism Among the Mestizo Population of the Peruvian Amazon* and, with Pablo Amaringo, *Ayahuasca Visions: The Religious Iconography of a Peruvian Shaman.* He is also co-editor—with Steven F. White—of *Ayahuasca Reader: Encounters with the Amazon's Sacred Vine.* In 1986 with Pablo Amaringo he co-founded the Usko-Ayar Amazonian School of Painting of Pucallpa, Peru, and served as its director of International Exhibitions until 1994. He was professor of anthropology at the Federal University of Santa Catarina, Brazil, from 1994 to 1998, has lectured about Amazonian shamanism and modified states of consciousness worldwide, and has curated exhibitions of visionary art in several countries. He is the director of Wasiwaska: Research Centre for the Study of Psycho integrator Plants, Visionary Art and Consciousness in Florianópolis, southern Brazil (its website is www.wasiwasiwaska. org.). He is also a senior lecturer at the Swedish School of Economics and Business Administration in Helsinki, Finland.

Dennis McKenna, Ph.D., is a renowned ethnopharmacologist who co-authored, with his brother Terence, a classic in the field of psychedelic literature, *The Invisible Landscape,* which recounted their wild adventures in pursuit of Amazonian hallucinogens in 1971. He co-stars in Terence's later book, *True Hallucinations,* which further describes that fateful trip. Dennis earned his Master's degree in botany at the University of Hawaii in 1979 and his Doctorate in botanical sciences in 1984 from the University of British Columbia. In the early 1990s he held positions at Shaman Pharmaceuticals (Director of Ethnopharmacology) and the Aveda Corporation (Senior Research Pharmacognosist). In 1998 Dennis co-founded the nonprofit Institute for Natural Products Research (INPR) to promote research and scientific education with respect to botanical medicines and other natural medicines. Dennis is also a founding board member of the Heffter Research Institute and serves on the advisory board of the American Botanical Council. He has served as a board member for Botanical Dimensions and he is the editor-in-chief of *The Natural Dietary Supplements Pocket Reference* and *Botanical Medicines: The Desk Reference for Major Herbal Supplements*. He is also the author of countless scientific papers.

Terence McKenna (1946–2000) became the most well-known public defender and exponent of the value of visionary states obtained through the ingestion of sacred plants. He graduated from the University of California at Berkeley during the ferment of the 1960s and began to travel extensively in the Asian and "New World" tropics, finally gravitating to his life-changing encounter with the shamanism and ethno-medicine of the Amazon Basin. He and his brother Dennis co-wrote a book about their wild adventures in the Amazon, *The Invisible Landscape*. Terence then went on to write two more books about the impact of psychotropic plants on human culture and evolution: *Food of the Gods* and *The Archaic Revival*. His other books include *True Hallucinations* and *Trialogues at the Edge of the West,* a record of his discussions with mathematician Ralph Abraham and British biologist Rupert Sheldrake.

John Mohawk, Ph.D. (1945–2006), a Turtle Clan Seneca, was in the late 1960s and early 1970s the editor of *Akwesasne Notes*, the largest Indian publication in the United States at that time and a crucially important, groundbreaking forum for North American indigenous struggles. John was an assistant professor of American Studies at the State University of New York at Buffalo and on the board of the Collective Heritage Institute (CHI, the parent organization of the renowned annual Bioneers Conference) until his death in December 2006. He was also a farmer who helped develop food products based on white corn and other traditional Iroquois foods. The author of several books, including *Utopian Legacies*—a fascinating look at some of Western thought's problematic dimensions— John was a uniquely original thinker on a wide range of topics.

Ethan Nadelmann, Ph.D., is one of the world's most high-profile critics and commentators on U.S. and international drug control policies. After serving as assistant professor of politics and public affairs at the Woodrow Wilson School of Public and International Affairs at Princeton, he founded a leading drug policy institute, The Lindesmith Center, with the support of George Soros. He now serves as the executive director of the Drug Policy Alliance, the nation's leading drug policy reform organization. His cogent and eloquent writings on drug policy have appeared in numerous scholarly and mainstream journals and publications, from *National Review* to *Rolling Stone*. He is the author of *Cops Across Borders: The Internationalization of U.S. Criminal Law Enforcement* and he co-edited *Psychoactive Drugs and Harm Reduction: From Faith to Science*. He speaks throughout the world and has appeared on scores of radio and TV programs.

Jeremy Narby, Ph.D., a Swiss-based anthropologist and indigenous land rights activist who has spent several years in the Amazon basin, grew up in Canada and Switzerland. He studied history at the University of Canterbury, and obtained his Doctorate in anthropology from Stanford University. He is the author of several books, including *The Cosmic Ser-*

pent: DNA and the Origins of Knowledge and *Intelligence in Nature* as well as the co-editor (with Francis Huxley) of *Shamans Through Time: 500 Years on the Path to Knowledge.* Jeremy has worked for two decades with a very effective Swiss NGO (non-governmental organization) by the name of Nouvelle Planète, an organization which, among its many projects around the world, supports indigenous Amazonian people in their efforts to preserve their territories and cultures. Jeremy raises funds and develops programs designed to empower Amazonian groups in mapping and defending their ancestral lands, and has collaborated with them so that they can establish their own educational and cultural institutions. A very substantial portion of the indigenous lands now protected in the Peruvian Amazon is protected because of his efforts.

Dale Pendell is a poet, software engineer, and amateur yet accomplished ethno-botanist with a unique perspective on the study, use, and representation of the plant world, especially in the European tradition. He is the author of several books, including the massive trilogy *Pharmako/Poeia: Plant Powers, Poisons, and Herbcraft; Pharmako/Dynamis: A Guide for Adepts of the Poison Path—Excitantia and Empathogenica;* and *Pharmako/Gnosis.* He also wrote *Living With Barbarians: A Few Plant Poems.* All of his writings have radically broadened the discussion about the past and present uses of powerful plants.

Michael Pollan, currently Knight Professor of Journalism at the UC Berkeley Graduate School and director of its Knight Program in Science and Environmental Journalism, is a contributing writer to the *New York Times Magazine* and the author of four books: *The Botany of Desire: A Plant's-Eye View of the World, A Place of My Own, Second Nature,* and *Omnivore's Dilemma.* For many years he served as executive editor of *Harper's* magazine. His writing has won numerous awards, including the Reuters/World Conservation Union Global Award in Environmental Journalism, the James Beard Award, and the Genesis Award from the American Humane Association. A dedicated gardener, Michael has articulated a brilliantly creative view of the relationships between plants and humans and the mutual dependence that has evolved between us and several of the key species we rely upon for food, medicine, altered consciousness, and aesthetic fulfillment.

Edison Saraiva, M.D., is a Brazilian physician and homeopathic doctor. A specialist in eco-toxicology and nutrition, he worked with the Brazilian Ministry of the Interior in the northwest Amazon for many years and is a long-time member of the União do Vegetal Church, which uses ayahuasca as a sacrament.

Florencio Siquera de Carvalho is a humble man with very little formal education who has lived in the Amazon much of his life. He is an important spiritual teacher in the União do Vegetal Church.

Paul Stamets is a renowned mycologist; certainly no one has done more to explore and share the remarkable nutritional, medicinal, and entheogenic value of mushrooms. He is the president of Fungi Perfecti, a mail-order business supplying cultures, equipment, and mycotechnologies to mushroom cultivators throughout the world. He has written six books including *Growing Gourmet and Medicinal Mushrooms, The Mushroom Cultivator, Psilocybin Mushrooms of the World, Mycelium Running: How Mushrooms Can Help Save the World,* and many articles and scholarly papers. His passion is to preserve, protect, and clone—from the old-growth forests of the Pacific Northwest—as many ancestral strains of mushrooms as possible.

Michael Stewartt, the founder of Wild Angels—a New Mexico-based group working on land and forest conservation in the United States, Mexico, and Central and South America—is an internationally recognized leader in environmental restoration. Prior to creating Wild Angels, Michael founded the airborne environmental education organization Lighthawk, and he served as that organization's executive director and chief pilot for fourteen years. His work has been honored by the governments of Chile, Belize, and Costa Rica; by corporations such as Chevron (from whom he received their Conservation Award); and by many conservation organizations, including the National Wildlife Federation and the Southern Utah Wilderness Alliance. He also has much experience in South America and, as a pilot trained to detect environmental destruction, he has witnessed—as few others have—the devastation visited upon the ecosystems and peasantry of that continent by our demented war on drugs.

Jane Straight is the proprietress of Allies, a small mail-order plant and seed business specializing in rare ethno-botanicals. Deeply passionate about plants, Jane has worked for many years to empower young people with honest information about drugs. She has advocated in her community and beyond for the development of healthy, non-punitive drug education curriculums and programs that include student interaction and community support and assistance.

Andrew Weil, M.D., trained at Harvard Medical School and went on to become the most famous pioneer of holistic/integrative medicine on the planet. He has impeccable mainstream medical credentials, but he also has a Harvard degree in botany and, like Wade Davis, was a student of the amazing pioneer in ethnobotany, Richard Evans Schultes. Andrew worked for thirteen years on the research staff of the Harvard Botanical Museum and traveled extensively throughout the Americas and Africa in the 1970s, studying indigenous and folk medical healing traditions. He has become famous for his books on natural medicine: *Natural Health, Natural Medicine; Spontaneous Healing; Eight Weeks to Optimum Health;* and *Eating Well for Optimum Health,* but he has also long been one of the sanest voices on drug use in our culture—promoting his views in controversial books such as *The Natural Mind* and *From Chocolate to Morphine.*

BOOKS OF RELATED INTEREST

Plant Spirit Shamanism
Traditional Techniques for Healing the Soul
by Ross Heaven and Howard G. Charing
Foreword by Pablo Amaringo

The Encyclopedia of Psychoactive Plants
Ethnopharmacology and Its Applications
by Christian Rätsch
Foreword by Albert Hofmann

Plants of the Gods
Their Sacred, Healing, and Hallucinogenic Powers
by Richard Evans Schultes, Albert Hofmann, and Christian Rätsch

Drugs of the Dreaming
Oneirogens: Salvia divinorum and Other Dream Enhancing Plants
by Gianluca Toro and Benjamin Thomas
Foreword by Jonathan Ott

Sacred Vine of Spirits: Ayahuasca
Edited by Ralph Metzner

Sacred Mushroom of Visions: Teonanácatl
A Sourcebook on the Psilocybin Mushroom
Edited by Ralph Metzner

DMT: The Spirit Molecule
*A Doctor's Revolutionary Research into the Biology of
Near-Death and Mystical Experiences*
by Rick Strassman, M.D.

The Secret Teachings of Plants
The Intelligence of the Heart in the Direct Perception of Nature
by Stephen Harrod Buhner

Inner Traditions • Bear & Company
P.O. Box 388
Rochester, VT 05767
1-800-246-8648
www.InnerTraditions.com

Or contact your local bookseller